Asylum Seekers, Social Work and Racism

DOI: 10.1057/9781137415042.0001

Other Palgrave Pivot titles

Michael Huxley: **The Dancer's World, 1920–1945: Modern Dancers and Their Practices Reconsidered**

Michael Longo and Philomena Murray: **Europe's Legitimacy Crisis: From Causes to Solutions**

Mark Lauchs, Andy Bain and Peter Bell: **Outlaw Motorcycle Gangs: A Theoretical Perspective**

Majid Yar: **Crime and the Imaginary of Disaster: Post-Apocalyptic Fictions and the Crisis of Social Order**

Sharon Hayes and Samantha Jeffries: **Romantic Terrorism: An Auto-Ethnography of Domestic Violence, Victimization and Survival**

Gideon Maas and Paul Jones: **Systemic Entrepreneurship: Contemporary Issues and Case Studies**

Surja Datta and Neil Oschlag-Michael: **Understanding and Managing IT Outsourcing: A Partnership Approach**

Keiichi Kubota and Hitoshi Takehara: **Reform and Price Discovery at the Tokyo Stock Exchange: From 1990 to 2012**

Emanuele Rossi and Rok Stepic: **Infrastructure Project Finance and Project Bonds in Europe**

Annalisa Furia: **The Foreign Aid Regime: Gift-Giving, States and Global Dis/Order**

C. J. T. Talar and Lawrence F. Barmann (editors): **Roman Catholic Modernists Confront the Great War**

Bernard Kelly: **Military Internees, Prisoners of War and the Irish State during the Second World War**

James Raven: **Lost Mansions: Essays on the Destruction of the Country House**

Luigino Bruni: **A Lexicon of Social Well-Being**

Michael Byron: **Submission and Subjection in Leviathan: Good Subjects in the Hobbesian Commonwealth**

Andrew Szanajda: **The Allies and the German Problem, 1941–1949: From Cooperation to Alternative Settlement**

Joseph E. Stiglitz and Refet S. Gürkaynak: **Taming Capital Flows: Capital Account Management in an Era of Globalization**

Steffen Mau: **Inequality, Marketization and the Majority Class: Why Did the European Middle Classes Accept Neo-Liberalism?**

Amelia Lambelet and Raphael Berthele: **Age and Foreign Language Learning in School**

Justin Robertson: **Localizing Global Finance: The Rise of Western-Style Private Equity in China**

DOI: 10.1057/9781137415042.0001

palgrave▸pivot

Asylum Seekers, Social Work and Racism

Shepard Masocha
Lecturer of Social Work, University of South Australia

DOI: 10.1057/9781137415042.0001

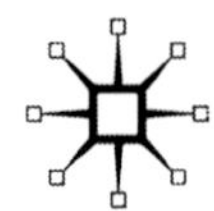

First published 2015 by
PALGRAVE MACMILLAN

Palgrave Macmillan in the UK is an imprint of Macmillan Publishers Limited, registered in England, company number 785998, of Houndmills, Basingstoke, Hampshire RG21 6XS.

Palgrave Macmillan in the US is a division of St Martin's Press LLC, 175 Fifth Avenue, New York, NY 10010.

Palgrave Macmillan is the global academic imprint of the above companies and has companies and representatives throughout the world.

Palgrave® and Macmillan® are registered trademarks in the United States, the United Kingdom, Europe and other countries.

ISBN: 978–1–137–41505–9 EPUB
ISBN: 978–1–137–41504–2 PDF
ISBN: 978–1–137–41503–5 Hardback

A catalogue record for this book is available from the British Library.

A catalog record for this book is available from the Library of Congress.

www.palgrave.com/pivot

DOI: 10.1057/9781137415042

In memoriam
Chance Masocha
(1978–2009)

DOI: 10.1057/9781137415042.0001

Contents

DOI: 10.1057/9781137415042.0001

Acknowledgements

I am forever indebted to my mum and dad who realised the value of education and decided to go without some of life's most basic comforts so that I could have a decent education; I hope one day your sacrifices will pay off.

This book would not have been possible without the intellectual guidance and support from a number of people. I am greatly indebted to Dr Murray K. Simpson, Dr Fernando Lannes Fernandes and the social work academic team at the University of Kent. As will become evident, Michel Foucault, Jonathan Potter and Margaret Wetherell significantly influenced my thinking. As always, I take full responsibility for all errors, omissions, misreadings and any other inaccuracies contained herein.

This achievement would not have been possible at all without the unwavering love and support from my wife, Sibongile, and my three daughters, Amanda, Kudzaishe and Zoe. A special thanks goes to all social work professionals who took time off their incredibly busy schedules to participate in interviews; thank you. I would like to thank my brothers and sisters for helping me believe at an early stage of my life that I could achieve anything if I want to; izvi hazvisi zvangu ndenga ndezvedu tose!

Lastly I would like to thank SAGE for permission to re-use as Chapter 5 of this book my article entitled 'Construction of the "other" in social work discourses of asylum seekers', *Journal of Social Work*, 2014, doi:10.1177/1468017314549502.

Terms and Definitions

Asylum seekers

Article 1(A)(2) of the Convention, as amended by the 1967 New York Protocol, provides the following definition of a refugee as a person:

> owing to well-founded fear of being persecuted for reasons of religion, nationality, membership of a particular social group or political opinion, is outside the country of his nationality and is unable, or owing to such fear is unwilling to avail himself of the protection of that country; or who, not having a nationality and being outside the country of his former habitual residence as a result of such events, is unable or, owing to such fear, is unwilling to return to it. (UNHCR, 2010b, p. 14)

The term refugee is widely used to describe displaced people all over the world. However, within the UK, a person is described as a refugee only after the Home Office has accepted their claim. Section 94(1) of the Immigration and Asylum Act (1999) defines an asylum seeker as follows: 'a person who is not under 18 and has made a claim for asylum which has been recorded by the Secretary of State but which has not been determined.' In other words, an asylum seeker is someone who has lodged a formal application for asylum/protection in the UK, and is awaiting a decision.

DOI: 10.1057/9781137415042.0003

Introduction

Abstract: *This section highlights the complexities and ever-shifting contours of racist discourse. It introduces xenoracism as a theoretical framework for understanding the ways in which asylum seekers are constructed. This section also discusses the theoretical and methodological approaches adopted in this book. The book uses a strand of discourse analysis called discursive psychology (Potter and Wetherell, 1987, Wetherell et al., 2001) as a theoretical and methodological approach to analyse spoken and written language as it is used to enact social and cultural perspectives as well as social identities.*

Masocha, Shepard. *Asylum Seekers, Social Work and Racism*. Basingstoke: Palgrave Macmillan, 2015.
DOI: 10.1057/9781137415042.0004.

Although the UK has witnessed a marked decline in the numbers of asylum applications from a peak of 84,130 in 2002 to 23,507 in 2013 (Refugee Council, 2014), the issue of asylum remains highly emotive and politically charged. At a practice level, social work professionals are faced with the difficult challenge of providing services that are both sensitive and responsive to the needs of an increasingly culturally plural service user group. This imperative of providing a culturally sensitive and responsive service is already widely acknowledged within social work, as evidenced by the centrality of anti-racist (Lavalette and Penketh, 2014, Singh, 2014, Dominelli, 1988) and anti-oppressive perspectives (Darylmple and Burke, 2006, Dominelli, 2008) within social work discourse. Studies undertaken on social work practice with asylum seekers (Collett, 2004, Humphries, 2004a, 2004b, Jordan, 2000, Sales, 2002a, Hayes, 2009, 2013) largely focus on asylum policies and their implications on practice. This approach is certainly valid and has resulted in a growing body of informative research, which has significantly led to the emergence of a critical social work perspective for understanding pertinent issues of ethnic diversity and social inclusion.

This book argues that social work continues to draw heavily on outdated views and definitions of racism that are largely based on skin colour and biological categorisation. This reliance on such narrow definitions of racism does not take into cognisance how the late 20th century witnessed the emergence of 'new racism' (Barker, 1981a). This book introduces social work to the concept of xenoracism (Sivanandan, 2001) in the hope that this will challenge the outdated concepts of racism that still pervade social work. It is anticipated that the book will result in both students and practitioners becoming more aware of the ever-shifting parameters of exclusionary discourses. More significantly, this book offers an additional perspective that will provide social work with a more in-depth understanding of how current social policy drivers and social policy itself are permeated by xenoracism. This is achieved primarily through an analysis of media and parliamentary discourses as well as social work professionals' discourses on asylum seeking. Such a critical understanding of the centrality of xenoracism in the construction of asylum seekers at these levels will enable social work to be more aware of the inevitable tensions and ethical dilemmas that emerge between what social work stands for and the current social policies within which social work has to operate.

DOI: 10.1057/9781137415042.0004

An important aspect that has remained largely unexplored within social work research is how asylum seekers are constructed as a social group and crucially how this process, which is underpinned by xenoracism, plays a pivotal role in the ways in which social policies relating to asylum seekers are formulated (Masocha and Simpson, 2011b). The ways in which asylum seekers are constructed have a fundamental impact on the sets of welfare apparatus that are then invoked. This book posits that the construction of social policies relating to asylum seekers is inherently racist. It will be argued that social work's value system and concepts of anti-oppressive and anti-racist practices on their own are inadequate as a basis for understanding and countering prevailing racist policy frameworks for asylum seekers. As such, within social work, there is a lack of an in-depth understanding of the nature of the existing policy frameworks for asylum seekers. Yet such an understanding is of particular relevance to social work practice given that asylum seekers are generally not perceived as part of mainstream society (Sales, 2002b), but as an anomaly (Klocker and Dunn, 2003). There is also growing evidence of the differential treatment received by asylum seekers within mainstream welfare services (Harris, 2003, Masocha, 2008, Dumper et al., 2006). An analysis of the ways in which asylum seekers are constructed at various levels of society including at practice level is also of particular interest in the light of the concerns raised in research by Humphries (2004b) and Collett (2004) that social workers are unwittingly being complicit in fostering inequality, oppression and social exclusion, which as a profession they purport to challenge.

Xenoracism

The late 20th century saw biologically defined racism being increasingly replaced within public discourses by a more sophisticated and subtle form of racism. Barker (1981a) discusses the emergence of this new form of racism in the context of New Right politics in Britain in the 1970s. He argues that this new form of racism occurred without explicit references to Social Darwinist notions of biological superiority and inferiority. This 'new racism' is so subtle that it is very difficult to identify and can easily go unnoticed. This subtlety of the new racism has led Wetherell and Potter (1992) to conclude that analytic frameworks such as those predominantly used within social work to define racism in terms of

DOI: 10.1057/9781137415042.0004

beliefs of biological superiority and the use of overt derogatory racist language risk becoming obsolete. According to Potter and Wetherell:

> Even the relatively blatant fascist propaganda and blatant advocates of racism (such as Le Pen in France) have learnt to modify their discourse so that on some occasions racism can occur without biological categorisation and the more familiar paraphernalia of 'advanced' and 'primitive,' 'negative' and 'positive,' 'superior' and 'inferior' distinctions. Given this flexibility of the enemy, and the way debates move on, it seems sensible not to commit oneself to one exclusive characterisation of racist claims. There is a danger of being silenced when racist discourse continues to oppress but no longer meets the main characteristics of social scientific definitions of racism. (1992, pp. 71–72)

It is therefore important that social work develops an awareness of the multivalent nature of racism. Reducing racism to only supremacist variants of racism ignores the more significant contemporary ways in which racist sentiments are couched and articulated. Significantly, there is also a risk that not fully appreciating the insidious and subtle forms in which racist sentiments are currently packaged may even result in social work legitimating the various manifestations of the new racism, which may easily be accepted as 'common sense' and reasonable arguments as will be illustrated in this book.

It is important to emphasise that the 'new racism' is not necessarily benign when compared to the 'old racism,' which is anchored in biological categorisations. It needs to be noted that the 'new racism' categorises the 'racial' other as presenting a specific and significant threat to the 'racial' in-group (Hopkins et al., 1997). In relation to immigration discourse within the UK, implicitly 'races' are defined using terms and categories that illuminate the biological and/or cultural characteristics of the immigrants. For instance, almost invariably within anti immigration arguments, negative lexical items are allocated to the immigrants whose presence is portrayed as problematic and resulting in negative consequences for the nation. This notion of *otherness* that characterises the prevailing discourses of asylum seekers fits well with this new type of racism. As will be demonstrated in Chapters 1 and 2, the prevailing discourses of asylum are preoccupied with who should be included or excluded from mainstream British society. As Gilroy (1992, p. 45), notes the new racism is also 'Primarily concerned with mechanisms of inclusion and exclusion. It specifies who may belong legitimately and simultaneously advances reasons for the segregation or banishment of those

DOI: 10.1057/9781137415042.0004

whose origin, [or] sentiment of citizenship assigns them elsewhere'. It is therefore important that social work develops a critical understanding of how the main drivers of social policy and social policy itself are permeated by these subtle forms of racism.

The last three decades have witnessed a significant increase in the amount of discursive research on race, immigration and asylum seekers. For instance, van Dijk (1984, 1987, 1991, 1993 and 1997b) has researched extensively on the new racism in relation to immigration and asylum seekers and also illustrated how these trends in the Netherlands are also replicated in other countries including the UK. In their analysis of Austrian political discourse, Krzyzanowski and Wodak (2009, p. 3) note that the right wing political parties have 'refined their electoral programs under the rubric of national populist slogans and adopted more subtle forms of racism'. Effectively these parties have moved away from overt neo-fascist discourses. Thus, the exclusionary tendencies of their discourses are no longer articulated in overt racial terms, which for instance used to refer explicitly to biological or racial terms. Instead, social characteristics are deployed as discursive resources aimed at distinguishing citizens from the 'undesirable' immigrants who are depicted as 'not wanting to work', 'a drain on public resources' and 'not wanting to integrate'. In New Zealand, Wetherell and Potter (1992, p. 70) studied racism in terms of 'discourse which has the effect of categorising, allocating and discriminating between certain groups'. They studied the effects predominantly by applying a 'practical skepticism to the discourse of white New Zealanders in order, as title has it, to provide a map of racist language' (Hammersley, 2003, p. 803). Other parts of the developed western world have also seen the growth of similar research including Australia (Augoustinos et al., 2005, Augoustinos, 2001) and Belgium (Blommaert, 2005). The UK has also witnessed the growth of discursive research, which seeks to examine this new racism (Reeves, 1983, Fowler, 1991, van Dijk, 1997b, Billig, 1988, Lynn and Lea, 2003). The common denominator of this extensive body of discursive research is the analysis of contemporary racist discourse. The research focuses on how linguistic resources are deployed to achieve the high levels of subtlety that are characteristic of the new racism. The research also focuses on how the new racism is characterised by covert talk yet at the same time it successfully articulates exclusionary and oppressive views. The various studies cited above also share a consensus that in terms of its effects, the new racism is not different from the 'old fashioned' racism because it

DOI: 10.1057/9781137415042.0004

justifies and legitimates inequalities as well as exclusionary tendencies. As such, there is an imperative for social work to be aware and conversant in these aspects of new racism if the current concerns about social work's complicity in oppressive and racist practices are to be successfully addressed.

It is important to note here that the notion of the *emergence* of a 'new racism' should not be accepted at face value, as there is a danger of obscuring the historical continuities that are evident in the history of racist discourses. For instance, the subtle expression of racial prejudice, a defining characteristic of new racism (and xenoracism itself), did not begin in the second half of the 20th century. As Leach (2005) illustrates, the expression of racial prejudice in 'subtle', 'symbolic', indirect and covert ways existed well before the second half of the 20th century. It is also worth noting that the blatant expression of racism based on biological categorisation, which is characteristic of 'old fashioned' racism, continues to exist in contemporary British society. In fact, one can argue justifiably that there is nothing 'new' about the current ways of expressing racial prejudice that are being identified as 'new racism'. Therefore, the notion of a 'new racism' should not obscure these important historical continuities. Essentially, what happens in the second half of the 20th century is a decisive shift in the manner in which racist sentiments are couched. Predominantly, blatant expressions of racism have been replaced (but not eliminated) by a subtle variety known as 'modern' (McConahay, 1986), symbolic (Kinder and Sears, 1981) and 'new racism' (Barker, 1981b). As society has become increasingly intolerant of blatant expressions of racial prejudice, a *coded discourse* has emerged which is understandable through semantic and pragmatic cues to wider audiences (Reisigl and Wodak, 2001). The extent to which such discourses are coded is largely dependent upon the level of tolerance within individual European countries.

Moreover, as Cohen (1999, p. 9) argues, the notions of 'new' and 'old' racism reintroduce binaries that do not fully reflect 'the complexity of what is happening on the ground'. The discourses of asylum seeking are so complex and fluid in nature that the concept of new racism on its own as an analytical framework may not result in a comprehensive analysis. Talking about racism, whether traditional or new type, conjures up images of white people pitted against black people; yet, as Card et al. (2005) showed there was no strong difference in attitudes towards immigration between people from ethnic minorities and those from

DOI: 10.1057/9781137415042.0004

the majority ethnic group in the UK. In any case, the racial composition of asylum seekers within the UK militates against using skin colour as a basis for analysis as asylum seekers are made up of different races. For instance, some of the asylum seekers from Zimbabwe and Eastern Europe are white. This shows the complex nature of the discourses at play. Therefore, there is a need to extend the analytical framework by combining new racism with the concept of xenophobia to come up with a new framework, xenoracism. Sivanandan (2001, p. 2) defines xenoracism as:

> a racism that is not just directed at those with darker skins, from the former colonial territories, but the newer categories of the displaced, the dispossessed and the uprooted, who are beating at Western Europe's doors... It is racism, that cannot be colour-coded, directed as it is at poor whites as well, and is therefore passed off as xenophobia, a 'natural' fear of strangers. But in the way it degenerates and reifies people before segregating and/or expelling them, it is a xenophobia that bears all the marks of... racism. It is racism in substance, but 'xeno' in form.

Therefore, xenoracism is a form of 'new racism', which is coded to disguise a very strong opposition to immigrants – 'it is a coded language that avoids overtly racist terminology while tapping into the prejudices of key voters who fully understand the unstated idea expressed' (Fekete, 2014, p. 36). Therefore, in spite of the apparent absence of definable targeted 'races', the discourse effectively reproduces racism all the same. Xenoracism is mainly associated with a rhetorically managed type of prejudice aimed at the discrimination, exclusion and marginalisation of asylum seekers. Asylum seekers are targeted not so much as a result of their respective phenotypical features such as skin colour but largely on the basis of a perception of foreignness. Thus, this coded discourse is essentially discriminatory because it targets individuals on the basis of perceived identity, cultural and religious differences. Augoustinos and Every (2007, p. 124) have noted that 'social taboos against expressing racist sentiments have led to the development of discursive strategies that present negative views of out-groups as reasonable and justified while at the same time protecting the speakers from charges of racism and prejudice'. Therefore, the xenoracist discourse is strategically and rhetorically organised to deny racism. As such, the book demonstrates that xenoracism is a salient feature in the manner in which asylum seekers in the UK are perceived, constructed and subsequently treated particularly in

DOI: 10.1057/9781137415042.0004

media and parliamentary discourses. This has significant implications on how asylum seekers are perceived, constructed and treated within social work. This book therefore uses xenoracism as an analytical framework for understanding the underlying attitudes and motives behind the various ways in which asylum seekers are constructed and subsequently treated. The prefix 'xeno' comes from the Greek word *xenos* which means alien. In ancient times, this meant anyone that was not Greek and therefore regarded as a barbarian. However, paradoxically, perhaps, the word *xenos* also means guest. In fact, there is a running thread through 'European' society that carries these two meanings in tension, as will be noted particularly in Chapters 1 and 2.

A turn to language use

The book uses discursive psychology (Wetherell et al., 2001) as a methodology for understanding the various ways in which asylum seekers are constructed. The term discourse analysis is used in this book as a generic term that encompasses research that focuses on language in its social and cognitive contexts (Potter and Wetherell, 1987). Discourse analysis provides a comprehensive understanding of how the rules and strategies of language use influence subjectivities and interactions, and vice versa (van Dijk, 1997a). Discourse analysis is employed to examine the ways in which society talks and writes about asylum seekers. Discourse analysis provides the tools for understanding how language use influences beliefs and interaction, and vice versa, how aspects of interaction influence how people speak or how beliefs control language use and interaction (van Dijk, 1997a). Discourse analysis is employed to examine:

(a) The ways in which the media, politicians and social work professionals construct asylum seekers
(b) The kind of thinking and sense making that is behind the existing discourses and
(c) The various ideological functions that such discourses serve especially in relation to the reproduction of social inequalities.

The book highlights some of the interpretative repertoires that were employed in the construction of asylum seekers and the various linguistic resources that were deployed by social work professionals in their attempts to explain and legitimate their practice with this service user

DOI: 10.1057/9781137415042.0004

group. Potter and Wetherell (1987, p. 128) define interpretative repertoires as 'a lexicon or register of terms and metaphors drawn upon to characterise and evaluate actions and events'. According to Edley (2001, p. 198), interpretative repertoires serve as 'The building blocks of conversation, a range of linguistic resources that can be drawn upon and utilised in the course of everyday social interaction'. As such, they are the 'recurrently used terms for characterizing and evaluating actions, events and other phenomena' (Potter and Wetherell, 1987, p. 149).

Interpretative repertoires that are presented in this book were identified through closely reading and re-reading a number of times corpora of data until

> one begins to feel as though you've heard it all before. People seem to be taking similar lines or making the arguments as others previously interviewed ... Gradually one begins to recognise patterns across different people's talk, particular images, metaphors and figures of speech. (Edley, 2001, pp. 198–199)

The interpretative repertoires identified provide access to specific speaking and acting positions, which carry with them specific roles and rights. Each interpretative repertoire creates a specific subject position for the respondent and establishes a particular kind of relationship between the respondent and his/her asylum seeking service users. Within the context of this book, the term subject position is accepted as referring to '... "locations" within conversations' (Edley, 2001, p. 210) where individuals are reconstituted as subjects in relation to relevant ways of speaking about asylum seekers. One of the central concerns of this book is the examination of the ways in which asylum seekers are positioned in the oppositional discourses, the rights and obligations that emanate from the assumed subject positions and the concomitant responses to asylum seekers that these subject positions entail.

Discourse analysis was conducted on parts of a corpus of interview data collected from 25 social workers who at the time were working with asylum seeking service users. The context for this book was a Scottish local authority that did not have a formal arrangement with the UK Border Agency (UKBA) to provide services to asylum seekers. Existing research in Scotland focuses on those local authorities that have formal arrangements with the UKBA for the dispersal of asylum seekers (Barclay and Ferguson, 2002, Wren, 2007, Coole, 2002, Sim and Bowes, 2007). However, asylum seekers also present in other local authorities, which

DOI: 10.1057/9781137415042.0004

do not have formal arrangements with the Asylum Support Service, which in itself presents its own challenges in terms of welfare provision. This book is based on one such local authority and as such provides a context that has not been widely explored.

Respondents were selected through making formal requests via the local authority for practitioners willing to participate in the research and informed consent was sought from participants before the interviews. The social workers came from statutory teams including child protection, Looked After Children, family support, housing, adult services and hospital-based teams. Interviews centred on respondents' experiences of working with asylum seekers. The interviewees were encouraged to talk about their views on asylum seeking, the roles of the media and politicians in constituting knowledge about asylum seekers, their individual experiences of working with asylum seekers, what they made of existing service provisions, perceived barriers and ways of enhancing practice. The intention was to acquire an in-depth understanding of the social workers' narratives and linguistic repertoires. Semi-structured interviews allowed participants the space to construct accounts of their own practice that were not constrained by a premeditated framework. Open-ended and follow-up questions were used to 'generate interpretative contexts in interviews' (Potter and Wetherell, 1987, p. 164).

The interviews were audio taped and transcribed and the social workers' accounts were randomly coded Social Worker 1–25 as a way of ensuring anonymity. Given that the approach adopted for this book is one that views an interview as a conversational encounter rather than simply a research instrument to reveal objectively a set of beliefs or opinions, in transcribing the recorded interviews the researcher's questions were considered as equally important as the respondent's answers. The researcher's questions were treated as setting some of the functional contexts for the respondents' answers and were included in the final transcript as it was acknowledged that 'the linguistic nuance of the question is as important as the linguistic nuance of the answers' (Potter and Wetherell, 1987, p. 165). In order to ensure authenticity and quality of the data collected through interviews, the transcripts were shared with the respondents with a view to correcting any errors in transcription. Nvivo8 was initially used to code and organise the transcribed interview data. In doing so, key discursive themes that recurred across and within accounts were unearthed. Noting recurring words, phrases, metaphors and arguments resulted in the identification of patterns in the data.

DOI: 10.1057/9781137415042.0004

These patterns became the subject of a rigorous analysis to reveal *how* asylum seekers were constructed in practitioners' narratives, the *purpose* and *function* of such constructs.

The perspectives of Potter and Wetherell (1987) and Phillips and Jorgensen (2002) were drawn upon to ensure that the data analysis was rigorous. The paper uses a range of textual features in analysing and negotiating the interpretative voice. Triangulation of data was particularly helpful in making sense of the relationship between the texts; the inter-discursive relationship between texts, genres and discourses relating to asylum seekers; the social level (context of situation); and the broader social, political and historical contexts (Wodak and Meyer, 2001). Constantly switching between these various levels during the analysis as well as evaluating the findings from these perspectives ensured that the research process was both robust and rigorous.

Thus, the book explores the everyday practices of social work professionals with asylum seeking service users and the specific ways in which these professionals explain and legitimate their practice with this group of service users. A central concern of the book is an examination of the meaning making processes that social work professionals engage in by focusing primarily on how they construct asylum seeking service users. However, the book acknowledges that these meaning making processes are embedded in a much wider context. Although the book takes a *micro approach* in its analysis, since its primary interest is on social work professionals' discourses and how they made sense of their everyday practice with asylum seeking service users, there is a need to acknowledge that such local meanings were embedded in wider contexts and discourses. This is because everyday interactions within social work practice with asylum seekers are also influenced by general discourses in society. For instance, social workers are not immune from media and politicians' constructs of asylum seekers and other groups of immigrants. As such, social workers are to some extent influenced by the wider discourses in society, which can impact on their practice. Geertz (1973, p. 5) refers to these as the wider 'webs of significance'. The book illustrates that social work professionals have other frames of reference outside their professional discourses which they can draw on to make sense of asylum seeking service users. As a result, the book includes Chapters 1 and 2, which analyse the history of the immigration and asylum policies, and media and parliamentary discourses, respectively. These chapters focus on these *macro* discourses

DOI: 10.1057/9781137415042.0004

in order to put into perspective and enhance the understanding of the micro discourses.

The social workers' discourses are shown as being argumentatively organised and oriented to the media and politicians' discourses. In this respect, the book establishes an understanding of how asylum seekers were constructed by social work professionals as it pays particular attention to the ideological basis of such constructs. This book illuminates the fact that language is one of the central vehicles through which social work takes place. As such, the analysis of social work discourse in its own right as a topic of analysis is a legitimate area of social work research, which can lead to an in-depth and enhanced understanding of social work practice itself. By using discourse analysis as a methodology, this book provides an important additional perspective for understanding not only social work practice with asylum seekers, but also some of the concerns regarding the profession's complicity in racist and oppressive practice.

An appreciation of how social work professionals construct asylum seeking service users will not only result in a significantly better understanding of contemporary social work processes but also a greater appreciation of their historical roots. The crucial role that discourse occupies in the construction and reconstruction of the social work profession is very significant. For instance, written texts, spoken discourse and various forms of non-verbal communication have played a critical role in the historical construction of social work practice, and continue to play a key role on the ongoing reproduction and reshaping of social work practices. Yet, in spite of apparent significance of discourse, research using discourse analysis to examine the dynamics of professional social work discourses is quite new and largely undeveloped within mainstream social work research. Within social work research, these mundane activities remain largely unexplored except for few notable exceptions (Fook, 2002, Hall et al., 2006, Hall, 1997, Parton and O'Byrne, 2000) which draw attention to the important role discourse plays in mediating practice.

DOI: 10.1057/9781137415042.0004

1
Historical Overview

Abstract: *This chapter provides an overview of the history of UK immigration and asylum seeking policies from 1900 to 2014. It reviews the public and government's responses to immigration and asylum. The rationale for this chapter is to provide a historical and social context for the ensuing analysis of the ways social work professionals construct asylum seekers. The chapter provides resources that allow links to be made between discursive patterns and social consequences. It is argued that the history of UK immigration and asylum policies is underpinned by an overriding desire to exclude and expel the Other. A version of history is thus presented upon which this book anchors its main argument that British immigration and asylum policies are permeated by xenoracism. It is argued that the policies regulating the asylum system are intrinsically racist and oppressive yet social work has to operate within and at times even enforce these policies. The social and historical context provided in this chapter therefore serves to clarify and make explicit this fundamental argument.*

Masocha, Shepard. *Asylum Seekers, Social Work and Racism*. Basingstoke: Palgrave Macmillan, 2015.
DOI: 10.1057/9781137415042.0005.

Introduction

The purpose of this chapter is to provide an overview of the history of immigration and asylum seeking in the United Kingdom. It also reviews the public and government's responses to these phenomena. The rationale for this chapter is to provide a historical and social context for the ensuing analysis of the ways social work professionals construct asylum seekers. Providing such a context in this analysis of professional discourses is necessary because it enables the 'interpretation of the position of a story, account or version of events within a field of power relations' (Thompson, 1984; cited in Wetherell and Potter, 1992, p. 105). As such, this chapter provides a backcloth for the analysis of social workers' discourse of asylum seekers. It provides resources that allow links to be made between discursive patterns and social consequences. However, by using history in this way, there is also awareness within this book that versions of history can never be presented in a simplistic and neutral form. As such, the historical and social context provided in this chapter should not be regarded as separate from but rather as an integral part of the whole interpretative process of this book.

This chapter represents a version of history upon which this book anchors its argument that British immigration and asylum policies are permeated by xenoracism. The policies regulating the asylum system are intrinsically racist and oppressive yet social work has to operate within and at times even enforce these policies. The social and historical context provided in this chapter therefore serves to clarify and make explicit this fundamental argument.

British immigration policies 1905–1981

Current British immigration and asylum policies should be understood as an integral part of an on-going process of discrimination based on one's country of origin and race. This process can be traced back to the beginning of the 20th century. Before looking at British asylum policy there is a need to outline the development of British immigration policy. This is important because in many ways immigration policies influenced and indeed set the tone for asylum policy. It was not until the early 20th century that Britain had a developed immigration policy even though the Poor Laws regulated internal movements. Since then, the development

DOI: 10.1057/9781137415042.0005

of immigration policy has been dominated by three main themes. First, is the tendency to construct immigrants as inherently problematic – 'outsiders', a 'danger' and a 'plague'. Secondly, immigrants are perceived in terms of the 'burden' they would impose on public finances especially the welfare system. Thirdly, the demands of the British economy have also been quite influential in shaping immigration policy.

It should be noted that well before the first piece of legislation was passed specifically to deal with immigration, the notion of immigrants as the 'outsider', a 'danger' and a 'plague' was an entrenched way of constructing immigrants within official circles and parliamentary discourses. For instance, as early as 1 April 1901, the Conservative Party MP for Sheffield Central, Howard Vincent asked if the Secretary of State for the Home Department had looked into:

> into the effect of alien immigration in the East End of London, and to the report made by their Commissioners to the effect that considerable areas in St. George's-in-the-East and other adjacent districts are being denuded of Gentile population, the properties sold, and the old tenants replaced by immigrants paying abnormally high rents, and defraying the expense by taking in an improper number of lodgers. (Howard Vincent, *Hansard*, vol. 92, Col: 347–348, 1 April 1901)

This negative perception of immigrants was also echoed in the 1903 Royal Commission's report which characterised the 'aliens' as 'impoverished, destitute, deficient in cleanliness, liable to introduce infectious diseases, criminals, anarchists, prostitutes, caused overcrowding, and raised rents' (cited in, Hayes, 2002, p. 31). Therefore, immigrants were constructed in discourses prevailing at the time as socially deviant and a threat to the British society. Thus the race vs. nation discourse anchored in the *us* against *them* distinction was very much evident then and was indeed to be drawn upon to justify subsequent restrictive legislation with negative exclusionary outcomes for immigrants. In fact, this us/them bifurcation continues to the present day to frame discourses relating to immigration and asylum and is one of the main salient and enduring features of xenoracism.

These negative perceptions and exclusionary tendencies towards immigrants can clearly be conveyed through the analysis of the parliamentary debates prior to the enactment of the first British immigration law, the 1905 Aliens Act. This law was proposed and passed specifically to establish barriers to the entry of Jews who were fleeing persecution

DOI: 10.1057/9781137415042.0005

in Germany, Russia and Poland as well as the general increase in the levels of immigration. Within the parliamentary debates that ensued, immigrants were portrayed predominantly in a negative light. The following, rather long, extract from the then Secretary of State for the Home Department, Akers-Douglas, when introducing the bill to parliament, is very important in conveying a clear picture of the perceptions and attitudes that existed at the time:

> there is a certain class of *undesirable aliens* who are *not so welcome*, and whose *repatriation is very desirable*... Unfortunately, these aliens have a tendency to occupy very few centres in this country, and therefore their presence *creates great difficulty* in certain districts. Between a fourth and a fifth of the whole of the foreign population in this country are residing in four or five centres. The last return shows that, excluding the *large families*, which many of them have, something like 54,000 are residents in the borough of Stepney. There are other boroughs in London where *large numbers* have also taken up their habitation; and by their residence in these districts they have not only *displaced a large amount of labour*, but have also occupied a very large number of dwellings from which they have driven *the bonâ fide inhabitants*.... Not only have these aliens living in these districts *caused a great deal of overcrowding*, with all its evils, and a *displacement of British labour*, but I am sorry to say, from the information which has reached me at the Home Office, that the feeling which exists between these settlements of foreigners in London and the native population is becoming very strained, and is really *a very serious menace to the maintenance of law and order* in these districts. *This evil*, I am sorry to say, is not likely to diminish; and, indeed, *it is increasing*. The immigration has increased very largely in recent years.... Another point which I would ask the House very seriously to consider is that *the class of aliens which we get here is not the class of aliens which at all makes the best citizens*. It is the class excluded by the United States, and therefore it is fair to say that *we only get the refuse*.... (emphasis added; Aker-Douglas, House of Commons, *Hansard*, 29 March 1904, vol. 132, Col: 987–995)

The importance of the above extract from Aker-Douglas' speech lies not only in its explicitly xenoracist depiction of immigrants, but also in the fact that the type of reasoning and justification for calls for restrictive immigration policies that were given here have been built upon and further elaborated (and possibly sanitised and presented in much more subtle form) in contemporary discourses as will be demonstrated. Some of the xenoracist notions that have shaped immigration policy are highlighted in the above extract by way of italics. In fact the notions that are

DOI: 10.1057/9781137415042.0005

expressed here by Aker-Douglas of immigrants as undesirable Others; presenting difficulties to the host nation; having unnecessarily large families which put a strain on public resources; benefiting at the expense of the British citizens; a menace and socially deviant which makes them incapable of fully integrating into the British mainstream society and become bona fide citizens – continue to be present in contemporary discourses relating to immigration in general and asylum seekers in particular.

The result was the 1905 Aliens Act, which promulgated the power to prevent the landing of undesirable immigrants in the United Kingdom. The perceptions, attitudes and exclusionary tendencies towards immigrants that prevailed at the time were written into this piece of legislation. According to Section 3 of the Act, an immigrant was considered undesirable if he was unable to support himself, or suffered from an illness that would result in the government shouldering the cost of treatment, or had a criminal history. As Cohen (1996) and Hayes (2002) argue, these determinations became the forerunners to the ways in which immigration rules are invoked to prevent immigrants from entering the United Kingdom particularly those perceived as requiring recourse to public funds. In fact, the law made provision for the deportation of immigrants found to be in receipt of parochial relief within their first year of arrival. As Hayes (2002, p. 36) notes it becomes clear 'how in the initial operation of the first piece of legislation to control aliens, access to public money remains key, both at the point of entry and internally' and this remains a key consideration in determining immigration and asylum cases.

The rationale and underpinning ideology for a restrictive immigration regime was based on the purported need to keep Britain 'British.' In fact the influence of Social Darwinism was explicit in the parliamentary debates leading to this Act as well as the media's portrayal of immigrants at the time. There was the underlying belief that 'non-Britons' came low in the pecking order of 'races'. Jews were, therefore, perceived as a major threat of alien dilution of English blood. It is in this respect that British immigration policy can be characterised as underpinned by racist philosophies. Jews were defined in the following terms: 'the real enemy, the invader from the east, the ruffian, the oriental parasite' (Hayes, 2002, p. 32). These views, centred on a need to safeguard 'Britishness,' led to the emergence of a discourse of nation and nationhood, which continues to shape contemporary immigration and asylum policies.

DOI: 10.1057/9781137415042.0005

The post-World War II period saw a slight change in policy towards encouraging immigration from the Caribbean and the Asian sub-continent. The change in the direction of immigration policy was necessitated by the severe post war labour shortages that bedevilled the British economy, which was experiencing a boom. It should however be stressed that this change in the direction of immigration policy was not followed by changes in attitudes and perceptions of immigrants. Racist attitudes and perceptions of immigrants as the Outsider continued to be present in immigration discourse. For instance, concerns regarding what were perceived as the deleterious effects of immigration of Black people on the 'racial character of the English people' were being voiced in 1948. Carter et al. (1987) state that two days after the arrival of the *Empire Windrush* a letter was sent to the then Prime Minister, Clement Attlee, by 11 Labour MPs calling for the control of Black immigration, since: 'An influx of coloured people domiciled here is likely to impair the harmony, strength and cohesion of our public and social life and to cause discord and unhappiness among all concerned.' The *Empire Windrush* had arrived on 22 June 1948 carrying 492 Caribbean immigrants. Such calls for a more restrictive immigration policy increased especially in the aftermath of the Notting Hill disturbances in September 1958. For instance, Louth MP, Cyril Osborne, argued that Britain was faced with 'the urgent need for a restriction upon immigration into this country, particularly of coloured immigrants' (House of Commons, *Hansard*, 29 October 1958b, vol. 594, Col: 195) and in December 1958 he forwarded a motion in parliament urging 'Her Majesty's government to restrict the immigration of all people... who are unfit, idle or criminal' (House of Commons, *Hansard*, 5 December 1958a, vol. 596, Col: 1552); common descriptors for black people at the time.

In fact, once the economy began to shrink in the 1960s culminating in the 1970s economic recession, immigration discourse focused on the 'problem' of immigration. On 23 February 1960, Frank Tomney, MP for Hammersmith North, in his contribution towards the debate on the 1958 Notting Hill disturbances, explained the riots as attributable to the 'problem' of immigration which led to ' a slow simmering to boiling point extending over two years, finally erupting in the mob violence ...' (Frank Tomney, House of Commons, *Hansard*, 23 February 1960, vol. 618, Col: 332–333). Thus, the post Second World War period saw what was articulated as an additional 'problem' associated with the influx from the Commonwealth. Immigrants were portrayed as threatening race relations within the

DOI: 10.1057/9781137415042.0005

British society. In fact, this argument continues to be used in calls to restrict the levels of immigration in contemporary discourses.

While in opposition throughout the 1950s and during the time when the Commonwealth Immigrants Bill was being debated in parliament, the Labour Party opposed calls to restrict immigration arguing that such calls were based on racism. Labour MP for Smethwick, Gordon Walker, in his contribution in the debate on the proposed legislation argued that the Conservative Home Secretary had been

> revealed before us in his nakedness. He is an advocate now of a Bill which contains bare-faced, open race discrimination. He advocates a Bill into which race discrimination is now written – not only into its spirit and its practice, but into its very letter. (Gordon Walker, House of Commons, *Hansard*, 16 November 1961, vol. 649, Col: 706)

In spite of the Labour Party's objections, the Commonwealth Immigrants Act (1962) was passed.

It is also important to underline the fact that in spite of its consistent opposition to immigration controls and the Commonwealth Immigrants Bill in particular during the period when Labour Party was in opposition, it was doubtful that a Labour Government would repeal the existing legislation. According to Sivanandan (1982, p. 12) once the Commonwealth Immigrants Act (1962) had been passed, 'the Labour Party with its eye to the elections had begun to sidle out of its commitment'. In late 1963, the Labour Party accepted the necessity of immigration controls. This was significant as it signalled the beginning of the emergence of a consensus between the Labour and Conservative parties on this issue. Thereafter, increasingly within Labour circles, there emerged the tendency to depict immigration as a problem. For instance, in 1965 Baroness Asquith of Yanbury described the 'problem' of immigration as a 'flood' that had all along been 'pouring in before the General Election when the previous Government were in power ... Yet that flood was apparently neither detected nor corrected' (House of Commons, *Hansard*, 10 March 1965a, vol. 264, Col: 78). Therefore, what emerged clearly in the Labour Party's parliamentary discourse on immigration was the notion of the undesirability of Commonwealth immigration. This was accounted for in terms of the perceived problems associated with immigration especially in relation to community integration and race relations.

Furthermore, when the Labour Government came into power in 1964, it came under increasing pressure over immigration from extreme

DOI: 10.1057/9781137415042.0005

right members of the Conservative Party particularly MPs from the Midlands region. For instance Smethwick MP, Peter Griffiths had successfully campaigned against Labour candidate Gordon Walker in the 1964 elections on the slogan 'if you want a nigger for a neighbour vote Labour'. The Conservative Party supported a new bill sponsored by Louth MP, Cyril Osborne, which aimed at denying entry to immigrants from the Commonwealth with the exception of those with parents born in the United Kingdom. Although the bill was thrown out of parliament, within a few months the Labour Government introduced a White Paper aimed at amending the 1962 Act by proposing a reduction on Commonwealth immigration quotas from 20,800 to 8,500 annually (The Earl of Longford, House of Commons, *Hansard*, 02 August 1965b, vol. 269, Col: 23–24). The outcome was the Commonwealth Immigrants Act (1965), which had the effect of further curtailing immigration from Commonwealth countries.

During the rest of the 1960s, the debates focused on the presence of black people from the Commonwealth. Within this discourse, imageries were deployed that evoked links with disease, crime and costs to the nation. It is within that context that Cyril Osbourne, MP for Louth, requested the Secretary of State for Commonwealth Affairs to call on Commonwealth countries to 'reduce the flow of migrants into Great Britain, in order to prevent a demand for their total exclusion until such time as the social problems created by those already here have been solved' (House of Commons, *Hansard*, 2 July 1968, vol. 767, Col: 1280). This clearly demonstrates the extent to which immigrants from the Commonwealth were perceived as the problematic Other.

Attention in the parliamentary debates turned to the increasing number of Kenyan Asian immigrants who held British passports and as such were exempt from the provisions of the Commonwealth Immigrants Act (1965). As British citizens they were free to enter and remain in the United Kingdom without any restrictions. The government reacted to the growing political pressure by introducing the Commonwealth Immigrants Act (1968) which withdrew the automatic right of entry into the United Kingdom of Kenyan Asians. For the first time a distinction was made between citizens who were patrials, that is, those who possessed identifiable ancestors in the British Isles, and those who did not. It became clear that those who were identified as patrials were exclusively white. The concept of patriality was subsequently written into the Immigration Act (1971). This law set up parameters for any

DOI: 10.1057/9781137415042.0005

claim to British nationality. Under section 2 of the Act, a person has the right of abode in the United Kingdom if:

i. he is a citizen of the United Kingdom or Colonies and that citizenship is on the basis of birth, adoption, naturalisation or registration in the United Kingdom;
ii. he is born as a citizen of the United Kingdom or Colonies born to or legally adopted by a parent who has citizenship status at the time of birth or adoption. The parent's citizenship was also supposed to be based on birth, adoption or naturalisation;
iii. he is a citizen of the United Kingdom and Colonies who has been settled in the United Kingdom and had at that time been ordinarily resident there for a period of five years or more; or
iv. he is a Commonwealth citizen born to or legally adopted by a British parent who at the time of the birth or adoption had citizenship of the United Kingdom on the basis of birth in the United Kingdom.

This was followed by the British Nationality Act (1981), which effectively removed the automatic right to British citizenship through birth in the United Kingdom and linked citizenship to right of entry. According to Humphries (2004b, p. 97), 'A myth that has underpinned immigration from the start, of the British nation comprising a distinct race of people of common origin and descent, white and European, was again written in the 1981 Act.'

In the light of the foregoing analysis of the evolution of British immigration policy, it can be argued that these successive pieces of legislations amounted to racial discrimination through immigration controls given that British nationality was explicitly linked to birth and by default a specific race. The effects of these changes to British nationality laws were inter alia to restrict the eligibility for citizenship for immigrants who managed to enter and remain in the United Kingdom. The strict eligibility criteria also denied citizenship to some of those who were born in the United Kingdom. In view of the provisions of these Acts, it can also be argued that the provisions clearly favoured white Commonwealth citizens over blacks in the determination of eligibility for British citizenship. On that basis these provisions can be characterised as racist. It is in view of the underlying notions of nationalism and chauvinism that Humphries (2004b) has characterised British immigration policies as inherently racist. Research by Anderson (1983), Anthias and Yuval-Davis

DOI: 10.1057/9781137415042.0005

(1989) and Miles (1993) has demonstrated that immigration laws depend on the distinction between 'native' and 'foreigner', 'citizen' and 'alien', 'us' and 'them' and 'legal' and 'illegal' (cited in Humphries, 2004b, p. 97). Indeed, these distinctions based on the us/them bifurcation permeate through both Labour Party and Conservative Party parliamentary discourse on immigration as demonstrated. As a result of the manner (xenoracist) in which immigration policy has evolved over time, what has become entrenched is the tendency to perceive an immigrant as the foreigner, the outsider, the Other and 'a seething mass of refuse and filth' (Gerrad, 1971, cited in Humphries, 2004b, p. 98). Closely linked to this is the perceived urgent need to protect the 'small' British island by closing off borders. These salient features were indeed instrumental and assumed prominence in the evolution of British asylum policy, and are of particular relevance to social work practice with asylum seekers.

British asylum policy under the Conservatives

The asylum issue only took centre stage in political debates from the mid-1980s. Prior to that, the issue of the arrival at the time of what were for the most part European asylum seekers in the post war period was pushed in the background. However, the 1980s saw a major shift in terms of the race of asylum seekers. The decade witnessed a significant increase in the numbers of asylum seekers who were considered, in generic terms, as black people. Arrivals of asylum seekers from Sri Lanka (Tamils), Africa and other parts of the Third World began to increase substantially. In Britain, the number of asylum seekers rose slowly initially from 1,563 in 1979 to 4,811 in 1986 and 5,100 in 1988. However, in 1989 the numbers went up threefold to 15,500 (Kaye, 1994). As asylum seekers were becoming more 'visible' in terms of race and skin colour, hostility among the British public began to rise. Consequently, the asylum topic assumed prominence in media and political discourses. This paved the way for restrictive legislation specifically to deal with the 'problem' of asylum seekers.

In response to the increase in the numbers claiming asylum, in 1985 visa requirements were introduced for Sri Lankan nationals. This was the first time visa restrictions were placed on Commonwealth nationals (Kaye, 1994). This policy of restricting entry was further reinforced by section 1 of the Carriers Liability Act (1987), which imposed heavy

DOI: 10.1057/9781137415042.0005

fines of £1000 per person on all carriers found guilty of transporting people who did not possess sufficient and appropriate documentation. According to Cohen (1994), the effect of this piece of legislation was that it extended the immigration service to employees of airlines and shipping companies as the onus was on them to detect anyone travelling with false documents.

More importantly, a discourse of disbelief emerged which cast asylum seekers as not meeting the conditions of refugee status as they were 'clearly arriving to better themselves' as stated in the 1991 Conservative Party's Campaign Guide (cited in Kaye, 1994, p. 150). As such asylum seekers were constructed as economic migrants trying to circumvent immigration restrictions by posing as asylum seekers. As the perceived threat grew the distinction between refugees and economic immigrants became blurred within the Conservative's language especially in the media. The Conservative's portrayal of asylum seekers in this way had the effect of rendering more acceptable to the public, the government's stringent measures to control the increasing numbers of people seeking asylum.

Legislation that was passed to govern the system of asylum clearly was therefore clearly not inspired by the need to protect and uphold the human rights of the persecuted that are seeking refuge. Instead, it was clearly a question of addressing numbers and the need to appear as a 'strong' government imposing 'strong' controls to guard against the possibility of the arrival of large numbers of asylum seekers who were portrayed as abusing the system in a bid to avoid immigration controls. The increasing numbers of asylum seekers arriving at the end of the 1980s placed the issue of asylum high on the policy agenda and it was in this context that proposals were introduced for the first piece of legislation dealing specifically with the issue of asylum.

The political debates surrounding the first ever piece of legislation proposed to deal specifically with asylum, the Asylum Bill (1991), clearly demonstrate the centrality of deterrence within the emerging asylum regime. The then Home Secretary, Kenneth Baker argued that the Asylum Bill would reform the treatment of asylum seekers and speed up the procedures for the determination of their claims and he believed that 'the rapid rejection of a large number of unfounded claims and the early departure of those applicants from this country will play a major part in deterring further abuse of the process' thereby allowing the government to deal effectively with what he saw as the ever growing numbers of

DOI: 10.1057/9781137415042.0005

asylum applications (House of Commons, *Hansard*, 2 July 1991b, vol. 194, Col. 166–1667). The Home Secretary's view projected the core belief that still exists today that the majority of asylum seekers are 'bogus' claimants. Michael Shersby, MP for Uxbridge, drew the attention of parliament to the fact that 'many bogus asylum seekers are coming to Britain as part of a carefully planned racket' (House of Commons, *Hansard*, 2 July 1991a, vol. 194, Col: 173). It is this belief that has been the foundation for much of the subsequent pieces of legislation relating to asylum seekers. The Home Secretary also voiced his concern about how Britain would be 'swamped' unless there was a concerted effort to curb immigration into Britain and Europe as a whole.

By this time there were broad areas of consensus that were beginning to emerge between the two main political parties. For instance, parliamentary debates are replete with acknowledgements by Labour and the Conservatives regarding the UK's obligations under the Geneva Convention. Both parties also shared the view that this obligation was unequivocally towards 'genuine' asylum seekers as opposed to the predominantly 'bogus' claimants who were presenting themselves in the United Kingdom. As a consequence both parties actively participated in a discourse of disbelief, which shaped asylum policies. The Labour Party also shared the view that most of the asylum seekers were not genuine although the overall party's position was to oppose the bill and cast it as essentially a racist piece of legislation.

However, the legislation ran into problems in parliament due in part to the opposition of not so much the Labour Party but rather from the alliance of various human rights Non Governmental Organisations (NGOs), churches and legal establishment that successfully campaigned against the bill. The legislation was eventually dropped and replaced by Asylum and Immigration Appeals Bill in 1992. What emerged clearly from the whole debate surrounding the bill was that it rarely articulated issues specifically related to the unique status of asylum seekers. The debates hardly centred on the need to safeguard and promote asylum seekers' rights. There was no mention of a resettlement plan for refugees. Instead, the emphasis was on deterring people from seeking refuge in the United Kingdom and indeed subsequent pieces of legislation followed the same template. Of particular concern to social work is the fact that since the early 1990s, as will be demonstrated, 'a main plank of deterrence has been a progressive dismantling of social rights for all asylum seekers, removing them from the usual provisions of citizenship' (Cemlyn and

DOI: 10.1057/9781137415042.0005

Briskman, 2003, p. 165). Clearly, this process is firmly based on the long-standing historical relationship between immigration control and access to welfare provision.

The Asylum and Immigration Appeals Act (1993) was enacted as part of a package of measures to address the growing number of asylum applications. Although it granted an in-country right of appeal against negative decisions to asylum seekers, this was only limited to 48 hours. This contrasted sharply with the 10-day period on standard immigration cases. The law included an extension of the Carriers Liability Act (1987) through requiring airline companies to demand transit visas to ensure that transit passengers did not disembark in the United Kingdom and claim asylum. The law also introduced compulsory fingerprinting of all asylum seekers. The statutory duty of local authorities to provide social housing for asylum seekers, if they had any temporary accommodation available was curtailed. This clause signified the beginning of the onslaught on asylum seekers' entitlement to mainstream welfare services. The 1993 Act had an almost immediate statistical impact. Six months prior to the act, of the 13,335 decisions made by the Home Office, 86 per cent were granted either asylum or Exceptional Leave to Remain (ELR) and only 14 per cent were refused. Six months after enactment, only 28 per cent were granted either asylum or ELR and 72 per cent were rejected (Stevens, 1998). This signalled a hardening of attitude by the Home Office officials and a determination by government to disseminate the message that the United Kingdom was not a 'soft touch'.

In 1996 the government passed the Asylum and Immigration Act. The then Home Secretary, Michael Howard, justified the act by linking immigration with welfare provision and employment opportunities. He argued that the 'asylum procedures are increasingly abused... The present benefit rules are an open invitation to persons from abroad to make unfounded asylum claims' and 'the fact that those people can get jobs quite easily – at the expense of those who are entitled to live and work here – is one of the main reasons why the United Kingdom is such an attractive destination to asylum seekers' (House of Commons, *Hansard*, 20 November 1995b, vol. 267, Col: 336–337). This provided the rationale for the move towards a closer link between benefit regulation and immigration status. According to Morris (1998, p. 951) this development should also be seen as a 'key tactic in the development of internal controls, both as a basis for interagency cooperation and the means by which service providers can be encouraged to police migration'. Thereon,

DOI: 10.1057/9781137415042.0005

immigration status became a key reference point in the manner in which authorities related to immigrants. Access to mainstream welfare provision inter alia became subject to one's immigration status.

The overall impact of the Act was the exclusion of asylum seekers from social citizenship resulting in the creation of a society in which there are institutionalised inequalities among people within the same country. Thus the UK society consists of 'citizens' who also include convention refugees on the one hand, and 'xenos' who are mainly asylum seekers who have minimal rights. Even these minimal rights that the 'xenos' possess are not guaranteed and are constantly eroded by the successive governments as illustrated throughout this chapter.

Asylum and immigration policy under New Labour

When Labour came to power in 1997 hopes of an asylum policy driven by a quest for social justice rather than self-serving political interests were soon dashed. It became apparent that the asylum policy of the Labour government was not different from that of the Conservatives. The government published the White Paper, *Fairer, Faster and Firmer – A Modern Approach to Immigration and Asylum* (Home Office, 1998), which was a consultation document that formed the basis for an even more restrictive legislation; the Immigration and Asylum Act (1999). The White Paper advocated for 'fair, fast and firm immigration controls' (Home Office, 1998), which was very reminiscent of the previous Home Secretary, Michael Howard's speech three years earlier. Howard had argued for 'firm but fair' immigrations controls (House of Commons, *Hansard*, 20 November 1995a, vol. 267, Col: 335). Therefore, it is no surprise that the 1999 Act builds on Conservative Policies in the areas of pre-entry controls and welfare support for asylum seekers. The only difference between the two governments was the extent they went in disenfranchising asylum seekers. The fact that Labour made a policy U-turn on immigration and asylum was picked on by Conservative MP, Anne Widdecombe in her contribution to the proposal to further restrict employment of asylum seekers under section 8 of the 1996 Act, which Labour had previously vehemently opposed:

> I am intrigued. I should like to know why, from a position which was clear when the Conservative Government introduced that measure; to a position that was apparently equally clear immediately after the election when the

> Minister was responsible for these matters; to now when we come before the House, the hon. Gentleman has decided that section 8 is a good thing. I am delighted. I welcome any sinner who repenteth. (House of Commons, *Hansard*, 16 June 1999, vol.333, Col: 484)

In fact, while in power the Labour government indeed went further than the Conservatives in further restricting asylum. The Carriers Liability Act was further extended to incorporate trucking companies and train passenger services like the Eurostar. The reason for this according to the then Home Secretary, Jack Straw, was 'to reduce the large numbers of undocumented passengers using this route' (House of Commons, *Hansard*, 8 April, 1998, vol. 310, Col: 255). Ironically, while in opposition, the Labour party had also argued against pre-entry controls saying they could prevent genuine asylum seekers from leaving the country where they were being persecuted. However, once in power, if they had had their way the Labour Government would have completely taken asylum seekers out of the social security system and cash economy by introducing a cashless system based on food vouchers only redeemable at designated supermarkets. It was only due to the threats of rebellion from some Labour back benchers that Jack Straw was forced to introduce a cash component of £10 in order to ensure that the Bill got through its third reading (Bloch, 2000). Under the revised legislation, welfare support for asylum seekers was to be administered separately by a new organisation called the National Asylum Support Service (NASS). Adults and children were to be given £10 in cash and their remaining benefits in food vouchers. Under this legislation, the total value of the social security benefit income support was equivalent to 70 per cent of the social security benefit income support received by eligible British citizens. Additional in-kind support in the form of furniture and the payment of utility bills increased the value of the package to 90 per cent of the prevailing value of income support. The policy succeeded in stigmatising, degrading and humiliating asylum seekers leading to their total exclusion from society in line with government's ultimate objective of making seeking asylum as unattractive as possible to potential applicants.

The Act also introduced the compulsory dispersal of asylum seekers in an attempt to relieve the burden of provision from London (Home Office, 1998). It is also important to draw attention to Regulation 20 of the Asylum Support Regulations 2000, drawn under the Immigration and Asylum Act 1999, which severely curtailed asylum seekers' freedom of movement. According to the regulation, asylum seekers and their

DOI: 10.1057/9781137415042.0005

dependents in NASS accommodation could not be absent for more than seven consecutive days and nights and no more than 14 days and nights in any six months period without the approval of NASS. Asylum seekers were to be subject to unannounced visits from NASS and immigration officials. The police were also to keep weekly updates of asylum seekers in the area including names, addresses, date of birth, nationality, gender of principal applicant and number of dependents. Asylum seekers could also be required to report to designated police stations at stated time periods. The designated police stations were often far from where asylum seekers lived resulting in asylum seekers walking over long distances to report as their support is largely in kind as illustrated. Cohen (2002) has described the NASS scheme and the dispersal system as the creation of modern Poor Law. Clearly, the Labour Government was committed to continuing with the dual strategy of their Conservative predecessors of restricting entry to the United Kingdom and reducing social citizenship rights for asylum seekers.

The failure of the Act to deal effectively with the growing number of applications meant that the Labour Government had to introduce even more stringent entry requirements and tougher legislation. For instance, social services have been drawn in increasingly as part of the government's attempts to get a grip on the asylum system. The Nationality, Immigration and Asylum Act 2002 went further by drawing the entire state machinery into the surveillance process. Local Authorities were under an obligation to furnish at the request of the Home Office information of any resident in their area suspected of unlawful presence in the United Kingdom. Local Authorities were also obliged to report failed asylum seekers who tried to claim community care provision (Humphries, 2004b) and this inevitably committed social services to be inquisitors of immigration status and reporters to the Home Office (Cohen, 2003). Section 55 of the Act further restricted available support by allowing the state to deny any support in the form of housing and state benefits to asylum seekers who lodged their applications more than 72 hours after arrival. Voluntary organisations such as the Refugee Council and Shelter have documented the destitution that resulted from this policy (Refugee Council, 2004b, Dwyer and Brown, 2008). Section 4 of the 2002 Act provides support to asylum seekers whose applications have been declined but have hit hard times. However, this support is on condition that the asylum seekers undertake to return to their respective countries of origin when called upon. Section 4 supports is set at a much

DOI: 10.1057/9781137415042.0005

lower rate than income support benefit currently set at a minimum of £56.80 per week. Asylum seekers supported under section 95 of the 2002 Act receive £43.94 per week while section 4 support is set at £35.39 per week (Gower, 2013). This figure has not been increased since April 2011, which equates to a decrease in real terms. The inadequacy of section 4 support also needs to be understood within the context of the huge backlog the government has to contend with when it comes to deporting asylum seekers whose claims have been rejected. As a result what in principle is supposed to be short-term support in reality is long-term support. This effectively sentences asylum seekers on section 4 support to a life of abject poverty especially given that they are not entitled to other forms of state support that can ameliorate their conditions. The Home Office reported that there were 4,831 asylum seekers on section 4 support, which represented a significant increase from the figure of 2,757 reported in the previous year (Home Office, 2013a, 2014).

The Asylum and Immigration Act (2004) made it a criminal offence for a person to possess 'without reasonable excuse' an invalid document showing his/her identity and nationality when first interviewed by an immigration officer upon entering the country. It gave the immigration officer the power to arrest the individual. The impact of this provision was that it situated asylum seeking within a criminological framework rather than a humanitarian framework that is underpinned by a genuine desire to protect individuals fleeing persecution (Fekete, 2014). The Refugee Council has raised concern that this measure penalises asylum seekers who arrive 'without travel or identity documents in effect punishing refugees for behaving like refugees' (Refugee Council, 2004a, p. 3).

A significant proportion of asylum applications fail simply because of poor or non-existent legal representation, not because their cases are unfounded. With no legal representation and often speaking little or no English, many asylum seekers stand no chance at all. Most asylum seekers depend on legal aid. At the same time, there has been a marked decline in the number of competent legal aid solicitors. This is mainly due to the fact that the government has over the years cut the amount of legal aid available for asylum cases. These cuts meant that asylum seekers are increasingly finding it difficult to access good quality legal representation (Robins, 2011). Two of the UK's largest immigration advice charities, the Refugee and Migrant Justice (RMJ) and the Immigration Advisory Service (IAS) collapsed in 2010 and July 2011 respectively. The IAS attributed its decision to enter into

DOI: 10.1057/9781137415042.0005

administration to the £350 million cuts that were being made to the legal aid system (Medley, 2011). Solicitors have pointed out that it is impossible to do a decent job representing asylum seekers with such low levels of funding.

In spite of the fact that it prides itself in its so-called policies of race equality, the Labour government was also actively pursuing a series of negative and reactionary administrative measures designed solely to keep certain races out. This distinction is embodied in the Race Relations (Amendment) Act 2000. While the Act extended anti-discriminatory legislation into the public sector, it significantly excluded those officials who make decisions on immigration cases thereby allowing them to make blanket decisions on the basis of country of origin. This clause has been described by Hugo Young, a journalist, as 'The bluntest piece of state sponsored ethnic discrimination in 35 years' (Young, 2001). In fact, when presenting the Race Relations Bill to the House of Lords, Lord Bassam justified the exemptions on the grounds that, 'The operation of immigration necessarily and legitimately entails discrimination on the basis of their nationality' (Lords *Hansard*, 20 October 1999, Col: 1268). This leaves people who are subject to immigration control susceptible to various forms of discrimination. This is why Cohen (2002, p. 539) has argued that, 'immigration controls represent institutionalised racism in the clearest sense'.

Immigration and asylum under the Coalition Government

In sharp contrast with the 2001 and 2005 election campaigns, the Conservative Party's 2010 election manifesto did not articulate the party's usual anti-immigration stance. For instance, there was no mention of 'asylum' in the manifesto. The section of the manifesto which dealt with immigration was positively framed; *Attract the brightest and best to our country* (Conservative Party, 2010). It stated that 'immigration has enriched our country and we want to attract the brightest and best people who can make a real difference to our economic growth'. However, this change of tone, which is signified by the absence of the usual anti-immigration rhetoric that had become synonymous with Tory election campaigns, should be understood in the context of David Cameron's strategy to rebrand and reposition the

DOI: 10.1057/9781137415042.0005

party within the landscape of British politics. The absence of a strong anti-immigration rhetoric was largely a result of an awareness that 'taking too tough a line risked alienating many middle-class, "small-l liberal" voters that the Tories needed to win over but who still tended to see them as the nasty party' (Bale et al., 2011, pp. 398–399). Thus, Cameron shifted the debate away from the all too familiar battleground of the 'spectre' of the so-called 'bogus' asylum seekers towards a focus on the impact of immigration on the economy. The Conservatives also pledged to reform the student visa system, which they highlighted an area of concern – 'the biggest weakness in our border control' (Conservative Party, 2010, p. 21) – that was prone to abuse. They also promised to apply transitional control measures for future EU ascension countries. They also noted the need for measures to be put in place to ensure that new immigrants integrated into their new communities. They promised to introduce a language requirement for those planning to settle in the United Kingdom on spouse visas. The most significant pledge made was 'to take steps to take net migration back to the levels of the 1990s – tens of thousands per year, not hundreds of thousands' (Conservative Party, 2010). It can be argued that the Tories' measured message on asylum and immigration in the run up to the May 2010 elections was carefully packaged, and strategically positioned in such a way that it would not only broaden the party's appeal, but quite significantly it enabled Cameron to draw upon the issue of immigration to successfully show both the then Labour Government and the Liberal Democrats (whose popularity surged after the first televised election debate) as inept and/or out of touch with the desires of the British people in relation to immigration.

It is hardly debatable that the Coalition Government's immigration policy is a Tory policy. This is because Coalition Government's immigration policy is largely based on the Conservative Party's election manifesto as the Liberal Democrats' election pledges were largely discarded. According to Bale and Hampshire (2012, p. 91), 'All accounts of the formation of the coalition suggest that immigration was a non-negotiable "red-line" for the Tories and that the Lib Dems were made aware of this from the off.' The government's immigration policy as spelt out in *The Coalition: Our programme for government* (HM Government, 2010) clearly indicates the extent to which the immigration policy echoes the Conservatives' election manifesto. The government undertook to do the following during its term:

DOI: 10.1057/9781137415042.0005

- Introduce an annual limit on non-EU economic migrants.
- End detention of children for immigration purposes.
- Apply transition measures for any countries joining the EU member states in future.
- Introduce new measures to tackle abuse of the British immigration system in particular the student visa route.
- Explore ways to speed up processing of asylum claims (HM Government, 2010).

Clearly, the only notable Lib Dem input to the Coalition Government's immigration policy was the intention to end child detention, which became the Coalition's first pronouncement in relation to immigration reform. However, in practice, this has proved unworkable in the absence of a viable alternative. Children continue to be detained for immigration purposes despite Nick Clegg characterising it as 'simply a moral outrage' and calling for a 'sense of decency and liberty to the way we conduct ourselves' (House of Commons, 2010). The government maintains that children are detained as a last resort and the periods of detention are limited to 72 hours in a 'new style family friendly secure "pre-departure accommodation"' (Gower and Hawkins, 2013, p. 9).

Reforming the asylum system

When the Coalition Government came into power in 2010, the numbers of asylum seekers had started to decline. In its programme of government, the Coalition pledged to 'explore new ways to improve the current asylum system to speed up the processing of applications' (HM Government, 2010). In the summer of 2010, an Asylum Improvement Project was launched involving a series of initiatives that were aimed at establishing best ways of processing applications, which would lead to better decision making, efficiency and cost effectiveness. The published progress report on the Asylum Improvement Project (UKBA, 2011) noted significant improvement in the processing of applications in particular the 450,000 backlog of legacy cases had been reviewed; 60 per cent of the applications were receiving decisions within 30 days; an increase in the quality of decisions that were being made; and a reduction of £100 million in the cost of asylum support in the previous 12 months (UKBA, 2011).

DOI: 10.1057/9781137415042.0005

Dealing with illegal immigration

The Coalition Government has also employed tactics to deal with illegal migrants and failed asylum seekers, which have proved very controversial and divisive. One such tactic in July 2013 involved immigration officers who were wearing stab vests stopping non-white commuters at Walthamstow, Stratford, Cricklewood and Kensal Green tube stations and conducting immigration checks to find out if they are in the country legally. The Home Office was reported to be tweeting live updates and pictures of 'illegal immigrants' being arrested (Andreou, 2013, Batty, 2013). In addition to the important question of what an illegal immigrant looks like, this incident also raises questions about the use of racial profiling and the disproportionate use of the controversial stop and search powers on ethnic minorities. Equally significant is the fact that there is no legal requirement within the United Kingdom to carry ID cards yet the targeted individuals were asked to positively identify themselves as legally residing in the country.

Even more controversial was a pilot scheme that had been launched by the Home Office which involved vans carrying large billboards with the following message directed at illegal immigrants; '*In the UK illegally? Go home or face arrest?*' The vans were being driven around Barking and Dagenham, Ealing, Barnet, Hunslow, Brent and Redbridge. The campaign attracted a lot of attention and opposition. The JCWI characterised the campaign as xenophobic and argued that it could potentially engender mistrust and divisions within a culturally diverse society like London (JCWI, 2013). In fact, it is possible to make links between the messages displayed on the billboards with the repatriation 'Go Home' slogans of the National Front, British National Party and the English Defence League. It is also possible to argue that the timing of this campaign itself could have been meant to check the growing popularity of the UK Independence Party. Labour M.P Diane Abbot tweeted that the scheme was a 'Party political broadcast on behalf of the Tory Party directed at the UKIP voters.' The scheme was opposed from within the Coalition Government. Nick Clegg expressed his displeasure with the scheme arguing that it is 'not a clever way of dealing with this issue', while Vince Cable called it 'stupid' and 'offensive' (Merrick, 2013). Due to the growing opposition to the scheme, the Home Office withdrew the vans. The legality of the scheme was challenged. The Refugee Migrant Forum of East London (Ramifel) contemplated taking legal action against the

DOI: 10.1057/9781137415042.0005

Home Office. Clearly, questions have to be asked on whether the Home Office, and by extension the Coalition Government, had abrogated its positive duty under the Equality Act to eliminate discrimination and foster good community relations.

Setting the stage for restrictive legislation

The Coalition Government introduced an Immigration Bill, which according to the then Minister of Immigration, Mark Harper, was supposed to 'make it more difficult for illegal immigrants to live in the United Kingdom unlawfully and ensure that legal migrants make a fair contribution to our key public services' (House of Commons, 2013b). The government launched two consultations containing proposals for the bill, which were likely to have far reaching consequences for immigrants including asylum seekers. The government's proposals were adopted in the new Immigration Act 2014 without any significant changes. The lack of any strong opposition to the proposals can largely be attributed to the growing cross party consensus on immigration issues.

Chapter 2 of the Immigration Act (2014) introduced a health levy to most temporary migrants, to be paid at the point of making the visa application for entry or extension of stay. The stated rationale for this is to protect public services and the benefit system from undue pressures that may be placed upon them by migrants and in particular illegal immigrants such as failed asylum seekers. This means that only non-EEA nationals with indefinite leave to remain would have free access to the NHS and the rest of the categories being charged for the service. This represents a radical departure from the requirement that gives NHS access to people 'ordinarily resident', that is, 'living lawfully in the United Kingdom voluntarily and for settled purposes as part of the order of their life for the time being, whether they have an identifiable purpose for their residence here and whether the purpose has a sufficient degree of continuity to be described as "settled"' (House of Commons, 2013b, pp. 12–13).

From a social justice point of view, these provisions are quite problematic. The NHS is funded through general taxation, which temporary immigrants and to some extent undocumented immigrants are also subject to. Effectively, this amounts to double taxation. Furthermore, charging the primary care services means that those immigrants, for

DOI: 10.1057/9781137415042.0005

example, failed asylum seekers, illegal and destitute immigrants, without the means to pay for GP consultations will not be seen resulting in negative health outcomes which may have serious implications for public health. This is likely to hit hard specific groups of migrants such as migrants with no recourse to public funds, failed asylum seekers and illegal immigrants that already exist on the very margins of society.

Significantly this Act also aims to 'prohibit illegal migrants from renting accommodation in the UK' as part of the 'Government's wider drive to prevent illegal immigration ... by removing the means by which migrants live in the UK unlawfully' (House of Commons, 2013a, House of Commons, *Hansard*, 3 July 2013b, vol. 565, Col: 56WS). The rationale for this is to 'send a clear and strong deterrent message, both here and overseas, that there are clear and practical consequences to breaking the UK's immigration laws' (Home Office, 2013b, p. 3). These provisions were enacted despite opposition from landlords. Chapter 1 of the Immigration Act 2014 establishes a duty that compels all private landlords to conduct immigration checks on tenants before providing residential accommodation. Failure to carry out these mandatory checks will result in heavy civil penalties of up to £3,000, which have been modelled on the existing civil penalty scheme for employers (Home Office, 2013b). To comply with the Equality Act 2010, everyone including British citizens are required to provide documentation of their legal status in order to circumvent unequal treatment of immigrants on the basis of their race (and nationality), which are protected characteristics under this law. However, it is not clear how the government intends to monitor this in practice to ensure that immigrants are not discriminated. There is a risk here that the requirement might be selectively applied to only those individuals, who may be perceived by prospective landlords as immigrants. There is also a risk that due to the heavy civil penalties, landlords will begin to see particular ethnicities and nationalities as a risk to their business, and as such less preferred tenants. Therefore, this legislation may inadvertently foster and entrench the discrimination of immigrants in the United Kingdom. The potential effects of Immigration Act 2014 are not very different of those of the infamous signs *No Blacks, No Dogs, No Irish.*

Furthermore, the 'red tape' associated with the verification service and the very complexities of the UK visa regime that the landlords have to deal with should not be underestimated. In its response to the consultation document for this Act, Migrants' Rights Network argued,

DOI: 10.1057/9781137415042.0005

> The evidence from a similar duty introduced for employers in 2008 under the Immigration, Asylum and Nationality Act 2006 suggests that the complicated nature of immigration and the fear of 'getting it wrong' could lead to problems. In 2009, the MNR reported evidence of confusion amongst employers about new checking duties, and that new employers checks had resulted in discrimination against ethnic minority workers by employers, and even in some cases facilitated wider exploitation of undocumented workers. (Migrants' Rights Network, 2013)

In addition to potential discrimination against immigrants, this new law is likely to result in this group being vulnerable to exploitation by rogue landlords. Furthermore, immigrants are being compelled to hand over sensitive personal information without any guarantee about how that information will be processed or stored.

Overall, the new law is likely to result in high levels of destitution of undocumented migrants including failed asylum seekers. This will increase their levels of vulnerability, as they do not have any recourse to public funds. It is likely that social services and other welfare services are likely to work with increasing numbers of such service users given that one of the intentions of the proposals is to 'flush out' illegal immigrants. These sweeping changes being introduced by the Coalition Government are likely to further marginalise asylum seekers and other groups of immigrants and clearly they impinge negatively on their civil liberties. In particular, some of the measures that have been implemented to date have effectively limited access to justice; have been tantamount to an unfair targeting and treatment of ethnic minorities. Certainly, some of the measures have significantly undermined the protection and upholding of human rights standards, and could potentially undermine community relations. It is therefore clear that the Coalition Government's immigration policy so far, like previous successive governments, is largely underpinned by a discourse that seeks to exclude those who are deemed as undesirable. It is this imperative to exclude this targeted group that has led to the adoption of a very restrictive asylum and immigration policy. Some of the government's measures so far especially in relation to its targeting and treatment of illegal migrants clearly demonstrate the xeno-racism that drives some of these policies. Some of the injustices and in particular the breach of asylum seekers' human rights have been brought into focus by the death of Jimmy Mubenga in October 2010 when he was restrained by G4S guards while on a deportation flight from the United Kingdom. The coroner heavily criticised current deportation practices

DOI: 10.1057/9781137415042.0005

and noted that there was evidence of the prevalence of racism among guards who escorted deportees (Taylor, 2013). Furthermore, recent research by Women for Refugee Women (WRW) focusing on Yarl's Wood detention centre detailed the negative experiences of women locked up for immigration purposes and highlighted evidence of harassment, physical, emotional and sexual abuse that some of the detained women were subjected to (Girma et al., 2014).

Discussion

This chapter has highlighted that the history of UK immigration and asylum policies is underpinned by an overriding desire to exclude and expel the Other. The chapter also illuminates how these policies are by nature underpinned by xenoracism. What is also clear in this chapter is that there has been an emerging consensus on asylum and immigration across the political divide on the necessity to discriminate, exclude and expel targeted specific groups of immigrants throughout this historical epoch. The post-Thatcher era in particular has seen the main political parties gravitating towards a broad consensus on immigration and asylum issues. According to Smith (2008, p. 416), 'immigration has essentially become a valence issue rather than a position issue: the main parties agree on the broad policy parameters and compete only on the detail of policy and implementation'. For instance, Favell (2008, p. 416) demonstrates that while the Conservatives may have portrayed themselves as 'strong' on immigration, the reality is that mainstream political parties' policies have been similar in practice save for the rhetoric. As a result of the emerging consensus on immigration and the need to appear 'tough', a very restrictive legislative framework has been established and social workers not only have to work within it but at times have to enforce it. This chapter has also highlighted how Local Authorities under the Nationality, Immigration and Asylum Act 2002 have been drafted into the policing and surveillance of asylum seekers.

There are significant and wide-ranging implications for social work practice. The ways in which immigration and asylum policies have been crafted represents significant ethical dilemmas for social workers. Social workers find themselves having to exclude, solely on the basis of immigration status, vulnerable groups they should be working with. According to Humphries (2004, p. 95), this has happened 'because they

DOI: 10.1057/9781137415042.0005

[social workers] are imbued with an individualistic and unpoliticised view of "values" concerned with being non-discriminatory, anti-racist and anti-oppressive, they can persuade themselves that "anti-oppressive" means what they say it means'. The tendency is for social workers to convince themselves that they are acting in an anti-discriminatory way and take comfort in that. This 'unacceptable practice' (Humphries, 2004) will continue unless social workers build in a critical tradition to their practice which will enable them to deconstruct how asylum seekers are portrayed at different levels in society. As such, anti-oppressive perspectives on their own are inadequate in fostering a critical tradition to social work practice. Anti-oppressive perspectives tend to focus on the client-social worker relationship and the inherent power dynamics. Although, there is an acknowledgement of the impacts of the wider economic and socio-political contexts on the client-social worker relationship, the anti-oppressive perspectives do not go beyond that in their analysis. For instance, they do not focus on the actual ways in which the social policies that regulate the client–social worker relationship are constructed and rendered acceptable.

As a result, there is a need for social workers to have a critical understanding of how immigration and asylum policies have developed over time as well as question their own roles when working with immigrant service user groups. As Lavalette and Penketh (2014, p. 14) cogently argue, there is a need for social workers to

> adopt a critical gaze towards the shifting political debates that shape our world. They need to reject the simplistic, superficial and common-sense explanations that blame minority and marginalized groups and instead dig beneath the surface to uncover the real relationships that are shaping our unequal social world. This requires engaging constantly with political and social debates about a range of issues that create and recreate the world within which social work operates, and that creates the 'public causes' of so much of the 'private pain' that affects the lives of social work service users.

This will result in social work practice that is both reflexive and critically reflective thereby avoiding the pitfall of 'implementing racist policy initiatives whilst maintaining its unreflective, self deceiving "anti oppressive" belief systems' (Humphries, 2004b, p. 95). Having such an in-depth understanding of how and why asylum seekers and other groups of immigrants are cast in such negative ways, which are underpinned by xenoracism, will result in social workers becoming much more confident

DOI: 10.1057/9781137415042.0005

when working with this service user group. Such knowledge will also enhance the capacity of social work as a profession to engage effectively in debates around its roles and remit as well as advocating more effectively for social justice for this service user group.

DOI: 10.1057/9781137415042.0005

2
Asylum Seekers in Media and Parliamentary Discourses

Abstract: *This chapter explores how the marginalisation of asylum seekers and their subsequent exclusion from the mainstream welfare apparatus and wider society have been legitimated and normalised in media and parliamentary discourses. The media and parliamentary discourses make up part of the wider frames of reference that social work professionals draw on in addition to their specific professional discourses. The chapter demonstrates how asylum seekers are depicted as a problematic out-group, the Other. Interpretative repertoires used in anti-asylum seeking discourses are analysed with specific reference to the linguistic devices deployed in the construction of asylum seekers as an out-group. The chapter demonstrates that xenoracism is not inherent in asylum seeking discourses but rather it is an* effect *of using specific discursive resources and rhetorical devices. The deployment of such rhetorical devices achieves specific acts, which cast asylum seekers in a negative light without the deployment of an overtly racist discourse. It is in this sense that it is argued that these various interpretative repertoires bear all the hallmarks of the xenoracism in terms of the subtlety of the exclusionary tendencies.*

Masocha, Shepard. *Asylum Seekers, Social Work and Racism*. Basingstoke: Palgrave Macmillan, 2015.
DOI: 10.1057/9781137415042.0006.

 DOI: 10.1057/9781137415042.0006

Introduction

This chapter provides an insight into the various ways in which asylum seekers are portrayed in two significant areas: the media and parliamentary discourses. This will serve to make explicit how the marginalisation of asylum seekers and their subsequent exclusion from the mainstream welfare apparatus and wider society have been legitimated and normalised. The rationale for this chapter is to provide an appreciation of a wider context within which the construction of asylum seekers by social work professionals can be understood. This is because the media and parliamentary discourses make up part of the wider frames of reference that social work professionals draw on in addition to their specific professional discourses, in their attempts to engage meaningfully and make sense of asylum seeking service users.

Rhetorical othering

What emerges clearly from the discussion of asylum policies in the previous chapter is how asylum seekers are depicted as a problematic out-group. This marginalisation of asylum seekers is achieved through a process characterised by Riggins (1997) as rhetorical Othering. This process involves the stigmatisation of asylum seekers as a targeted social group as they are perceived as subversive, dangerous and in particular illegitimate. This portrayal of asylum seekers as an out-group is accomplished through the use of the oppositional binary *us v them,* which is characterised by the positive portrayal of groups and individuals who are subsumed under the first person plural pronouns *us* and *we*; and the simultaneous marginalisation of those groups that are designated as *they* and *them*. However, the dividing line between the oppositional binaries is one that is discursively drawn and redrawn as the ideas of *sameness* and *difference* as well as claims to ownership (*ours/theirs*) and group membership (*us/them*) are staked out and contested. Fowler (1991) examines the ways in which language structures encode an ideological viewpoint and the ways in which the media acts in perpetuating unequal power relations through these linguistic structures:

> In the Press, this ideology is the source of the 'consensual "we"' pronoun which is used often in editorials that claim to speak for 'the people'. How 'we' are supposed to behave is exemplified by the regular news reports of stories

DOI: 10.1057/9781137415042.0006

> which illustrate such qualities as fortitude, patriotism, sentiment, industry. But although consensus sounds like a liberal, humane and generous theory of social action and attitudes, in practice it breeds divisive and alienating attitudes, a dichotomous vision of 'us' and 'them'. In order to place a fence around 'us', the popular papers of the Right are obsessed with stories which cast 'them' in a bad light: trades unionists, socialist council leaders, teachers, blacks, social workers, rapists, homosexuals, etc., all become stigmatized 'groups', and are then somehow all lumped together and cast beyond the pale. (Fowler, 1991, p. 16)

The binaries that are established through using personal subjective pronouns are used to construct in-groups and out-groups. This also applies to the ways in which asylum seekers are portrayed in media and parliamentary discourses. In the case of portrayal of asylum seekers in media discourses, the use of inclusive pronouns *we/us* in addition to denoting an out-group simultaneously distances the reader/listener from *them*.

The use of the oppositional binary *us/them* is also important in framing the discourse of nations and nationhood which is a crucial component in the ways in which asylum seekers are constructed in media and parliamentary discourses. A recurring theme throughout this chapter is how the construction of asylum seekers in media and parliamentary discourses is framed as constituting a national crisis or a national invasion which has the potential to destabilise, undermine or threaten the British communities and their ways of life. Lynn and Lea (2003) have noted how in the United Kingdom, the notion of asylum seekers as presenting a national crisis tends to take centre stage of national debates during elections and in parliamentary debates. Hier and Greenberg (2002) have also reported similar trends in Canada. Billig (1995) analyses the deployment and effect of discourses of nation and nationhood in the speeches of US President Bush in the run up to the first Gulf War. This analysis is particularly informative and relevant as it demonstrates the significance of invoking the discourse of nation and nationhood as a linguistic strategy. Billig (1995) illustrates how President Bush worked up the discourses of nation and nationhood in his construction of the war to portray it as a perfectly legitimate and necessary defence of US *national* values; and the protection and defence of another *nation*'s sovereignty, that is Kuwait. He argues that this linguistic strategy was largely responsible for the huge popularity that the decision to go to war then enjoyed as evidenced by the increase in Bush's public opinion poll ratings and the

DOI: 10.1057/9781137415042.0006

increases in the sales of patriotic memorabilia, thereby demonstrating 'the speed with which the Western public can be mobilised for flag-waving warfare in the name of nationhood' (Billig, 1995, p. 2).

The idea of a nation has become a common sense and taken-for-granted notion that permeates through all aspects of daily life and everyday talk. Billig (1995) argues for instance that the idea of a nation is implicit in the use of pronouns such as *we* and *our*; for example, 'our' Prime Minister, 'our' sovereignty and 'our' country. This everyday 'banal' nationalism (Billig, 1995) which characterises everyday talk has become so entrenched and normalised in western societies to the extent that they have become invisible and unnoticeable. According to Billig (1995) it is on these ideological foundations of banal nationalism that the legitimating power of invoking the nation as a discursive resource can be revealed. Therefore, Billig (1995, p. 7) argues that, 'The popular reaction of support for the Gulf War in the United States cannot be understood by what happened in the moments of crisis. A banal, but far from benign, influence must have been routinely accomplished to make such readiness possible.' Therefore, this banality cannot be mistaken for being benign or harmless as it is hegemonic and caters to those within the *us* in-group (Sonwalker, 2005).

It is important to note that the *us/them* bifurcation is central to the notion of nationhood. As Nag (2001, cited in Sonwalker, 2005, p. 270) notes:

> Nations have always been concerned about 'us' as against 'them'. Nations are obsessed with 'self' and discriminate 'the other'. The construction of the national self has always been only vis-à-vis 'the other'. The basis of such construction is differentiation. The 'self' consisted of people who share common cultural characteristics and such commonalities could be measured by contrasting against those who do not. Thus construction of nationhood is a narcissistic practice while nation building is all about walls around the 'self' and distancing from 'the other'. (cited in Sonwalker, 2005, p. 270)

Therefore, the notion of nationhood is central in bringing together and mobilising people as a collective group and also distinguishing those who do not fall within the boundaries of the collective group. As such, one's national identity can be articulated as a basis for differentiating between who belongs and who is an 'outsider'. Reicher and Hopkins (2001) note how one's national identity can play an important role in the specific ways in which as the Other, one may to treated within the western world. They argue that the responses to the Other can be worked up in terms

DOI: 10.1057/9781137415042.0006

of the ways in which the Other is perceived to impact on the nation: 'If they enhance the national interest they are embraced; if they threaten the national interest they are to be rejected' (Reicher and Hopkins, 2001, p. 77). This is a recurrent and important theme in discourses relating to asylum seekers because various interlocutors frame their views this way as will be illustrated in this chapter.

Construction of asylum seekers in media discourses

Some sections of the British media have assumed a leading role in conveying and fostering negative perceptions of asylum seekers. According to Okitikpi and Amyer (2003, p. 216), discussions about asylum in Britain are characterised by '... negative attitudes, emotive language that depersonalises and criminalizes those seeking refuge'. Asylum seekers have been portrayed in the media as hordes of economic refugees, scroungers, bogus, fraudsters and parasites as will be demonstrated in this chapter. The media is a key political actor when it comes to framing the asylum debate. A European Human Rights Commission aptly summed up the negative roles of politicians and the media in influencing public perception of asylum seekers:

> The problems of xenophobia, racism, and discrimination persist and are particularly acute vis-à-vis asylum seekers and refugees. This is not only reflected in the xenophobic and intolerant coverage of these groups in the media, but also in the tone of the discourse resorted to by politicians in support of the adoption of and reinforcement of increasingly restrictive asylum laws. (Bloch and Schuster, 2002, p. 406)

Tabloid newspapers especially *The Sun* and *The Daily Mail* have been particularly venomous and negative in their portrayal of asylum seekers.

The case of Dover in 1999 clearly illustrates the power that the media is capable of exercising. From 1996 to 1999, an estimated 750 Roma asylum seekers from Slovakia and the Czech Republic were living in Dover. Their presence was sensationalised particularly in the *Dover Express* and the *Folkestone Express* which referred to thousands of asylum seekers flooding Dover and putting a strain on public resource. Asylum seekers were negatively portrayed as 'bootleggers' and 'scum of the earth', 'targeting our beloved coastline' (Maisokwadzo, 2004) and one of the papers calling on the people of Dover to reject asylum seekers whom it characterised

DOI: 10.1057/9781137415042.0006

as 'the back draft of a nation's human sewage' (*Dover Express*, 1 October 1998). The result was a marked increase in tensions and violence directed at asylum seekers. This culminated in a widely reported altercation between a group of asylum seekers and local youths in August 1999. The *Daily Mail* published an article sensationally entitled '*Handouts galore! Welcome to soft touch Britain's welfare paradise: Why life here for them is just like a lottery win*' (*Daily Mail*, 10 October 1997). It published the findings of its own research into Britain's immigration crisis under the headline '*The Good Life On Asylum Alley*' (*Daily Mail*, 6 October 1998). Dover was portrayed as a small town that was under threat from multitudes of foreigners who were not genuine asylum seekers but merely 'playing the asylum appeals process'. The *Daily Mail* also carried headlines such as '*The brutal crimes of asylum seekers*', which attributed the increase in crime in London to the presence of asylum seekers (Williams, 1998). Another article entitled '*Suburbia's Little Somalia*', accused asylum seekers from Somali, living in 'affluent, middle class Ealing ... thousands of miles away from the dusty plains of East Africa', of being involved in drugs and crime which were having a negative effect on the affluent neighbourhood (Goodwin, 1999).

Within contemporary discourses of cultural and community integration, food is normally held as a celebration of multiculturalism (Hage, 1997). On the contrary, asylum seekers' food habits have been a target of the media in its attempts to portray asylum seekers in a negative light. For instance, *The Sun* newspaper's article on 4 July 2003, entitled *Swan Bake*, drew on an old and archaic English law to accuse asylum seekers of stealing the 'Queen's swans for a barbeque'. This was followed up by another article in *The Daily Star* on 21 August 2003 under the headlines, *Asylum Seekers Eat Our Donkeys* and *Hands Off Our Asses* accusing asylum seekers of stealing some of the donkeys used for rides at Greenwich Royal Park; and within the article it was claimed that 'donkey meat is a speciality in some of the East African countries, including Somalia. And two areas near Greenwich – Woolwich and Thamesmead – have large numbers of Somalian (sic) asylum seekers'. The effect of constructing asylum seekers by referring to their food habits is that it depicts them as the Others who are not *like us*, and are a real threat to British ways of life. The other effect of these articles is that, asylum seekers are constructed as social deviants that do not belong, are out of place and polluting the host country's ways of life (Malkki, 1995) and the criminality of asylum seekers is a common thread that runs throughout these articles.

DOI: 10.1057/9781137415042.0006

Various tabloid reports seem to share the consensus that a threshold has been crossed and as such a violent reaction to the presence of asylum seekers in British communities such as Dover was somewhat understandable. It is important to point out that the vivid lexicalisation of asylum seekers is perhaps much more pronounced in tabloid newspapers as the above references suggest. However, broadsheet papers like *The Times* are subtler in the ways in which they insidiously construct asylum seekers. For instance, *The Times* mainly draws a negative picture of immigration and asylum policies by focusing on the 'victims' and insinuating the potential for racial tensions inside the United Kingdom if immigration continues unabated. *The Times* also pays particular attention to what it portrays as the government scandal on the immigration. However, it is less outspoken and dramatic in its accounts when compared with the representations of asylum seekers in tabloid newspapers.

The overall impact of these kinds of negative messages emanating from the media has been the depersonalisation, demonisation and victimisation of asylum seekers. The overall outcome of such predominantly negative media coverage of the asylum topic has been to foster and help popularise a disdain for those perceived as foreigners, which is synonymous with xenoracism. Therefore the media is certainly the thread that binds the discourse (Lynn and Lea, 2003) surrounding asylum seekers and indeed plays a pivotal role in the production and construction of particular forms of knowledge. In defining and categorising those perceived as the 'Other', the press employs various forms of imagery to project the visible difference, religious beliefs and language differences. Language can be an extremely powerful vehicle through which discrimination can be effected and sustained against particular groups of people. This is because,

> Language provides names for categories and also helps to set their boundaries and relationships; and discourse allows those names to be spoken and written frequently, so contributing to the apparent reality and currency of the categories. (Fowler, 1991, p. 94)

As a result the once 'morally untouchable category of political refugee' has been deconstructed (Cohen, 1994) by the media and replaced with the figure of the exploitative and criminal asylum seeker, who seeks to abuse 'soft touch' Britain (Conservative Party, 2001). This has made it possible for political parties to easily manufacture a moral and

DOI: 10.1057/9781137415042.0006

political consensus against asylum seekers through the use of language that removes the notion of legitimacy (Kaye, 1994).

Research studies by Fowler (1991); Hodge and Kress (1993); Fairclough (1995); and Fairclough and Wodak (1997) have discussed the ways in which the marginalised groups like asylum seekers are portrayed in the press. The focus of research by Bailey and Harindranath (2005) is the representation of asylum seekers in Australian news media and in the news programmes aired in the UK's BBC and Channel 4 stations. They analyse the Australian media's reaction to the sinking in Australian waters of an Indonesian ferry carrying asylum seekers. They argue that asylum seekers were depicted as *illegal immigrants*, and that these depictions played on the legal/illegal binary. These depictions were used to justify the need to strengthen national borders and invoked 'separatist discourses that clearly distinguish between the 'us' within the nation state and 'them', the outsider, the foreigner, the bogus refugee' (Bailey and Harindranath, 2005, p. 278). Therefore by drawing on the *us* versus *them* oppositional binary a nationalist discourse, which emphasises national security and nationhood, was invoked to marginalise the constructed out-group through the calls for more stringent asylum legislation. Drawing on a corpus of 27 editorials taken from newspapers in the immediate aftermath of 9/11, Achugar (2004) discusses the positive and negative representation of 'social actors' in those events with a particular focus on the negative representation of the Muslim as the Other. Elsewhere outside the United Kingdom, discursive research has also revealed similar roles of the media in shaping the discourse on asylum seekers. For instance, van Dijk (1991) provides a comparative analysis of the role of the press in the negative portrayal of asylum seekers in the Netherlands and the United Kingdom. In Australia, Pickering (2001, p. 169) examined the discourses in the print media and concluded that refugees and asylum seekers were 'routinely constructed not only as a 'problem', but also as a 'deviant' population in relation to the integrity of the state, race and disease'.

Construction of asylum seekers in parliamentary discourses

This construction of asylum seekers as the Other is further reaffirmed and formalised in parliamentary discourses. Politicians play a significant

DOI: 10.1057/9781137415042.0006

role in the definition of asylum seekers because they are the ones who make the crucial decisions on immigration, immigration restrictions as well as decisions relating to immigrants' eligibility for mainstream welfare provisions once they have been admitted into the country. According to van Dijk:

> If... elite groups... engage in discrimination against immigrants or minorities, the consequences are considerable: the 'Other' will not be allowed into the country in the first place, or they will not get a job, or they will not get promoted in their job, will not get decent housing, or the mass media or textbooks will spread negative stereotypes about them... the role of leading politicians, journalists, corporate managers, teachers, scholars, judges, police officers and bureaucrats, among others, is crucial for the (un)equal access to material or symbolic resources in society. (Every and Augoustinos, 2007b, p. 415–416)

As such politicians play a crucial role in the official definition and construction of asylum seekers. According to van Dijk (1997b) and Reeves (1983) politicians do not provide such definitions from scratch as they derive their information and beliefs in part from other elite sources such as the mass media, bureaucratic reports, academic research as well as their interaction with other elites. Although officially politicians are supposed to represent the wishes and views of their constituents in line with democratic norms and theory, in reality politicians' access to public opinion is marginal and at best indirect. According to van Dijk (1997b, p. 34):

> Popular resentment against immigration, such as in Western Europe, is filtered through the constructions or interpretations of popular reactions by journalists or other professionals. This means that both the media and the politicians are able to construct popular resentment as meaning what they please, for instance, as 'democratic' majority legitimation for the restriction of immigration or civil rights.

It also important to note that, the media and other elite discourses are also conversely influenced by the political discourses and the decisions made by politicians (van Dijk, 1991). Therefore, media and political discourses are products of complex inter-elite influences. As such, any analysis of political cognition and discourse should take into account such multiple influences and dependencies. Within this context, it is not only the power politicians can exercise in legislating and making policies that are crucial in the reproduction of existing inequalities. The influence that politicians can have on public discourses through the

DOI: 10.1057/9781137415042.0006

media and ultimately on the public opinion is also crucial (van Dijk, 1997b). Therefore given the far-reaching nature of their influence, an understanding of how asylum seekers are constructed in the media and parliamentary debates is necessary as social workers themselves are not immune to these discourses.

Interpretative repertoires in anti-asylum seeking discourses

An analysis of selected texts from newspaper articles and parliamentary debates in the House of Commons revealed a number of linguistic strategies that were employed in the construction of asylum seekers negatively. It should be noted that most of the contributions to parliamentary debates on asylum seekers are 'for the record' and are usually prepared well in advance. Given the current sensitivities around ethnic and racial affairs such institutional talk is highly self-controlled and carefully worded. In addition, given the existing legislation that prohibits discrimination and expression of racial hatred, politicians refrain from overt blatant expressions of prejudice. As a result, politicians resort to subtle and indirect ways of expressing prejudice, which are characteristic of xenoracism (Dovidio and Gaertner, 1986). In both media and parliamentary discourses, speakers maintain a subject position from which they strategically articulate their views and present them as reasonable. This is achieved through constructing their arguments in such a way that they undermine and rebut potential charges of prejudice, racism or xenophobia (Every and Augoustinos, 2007b). The following five interpretative repertoires that were identified in the analysis of the selected texts illustrate this:

(a) Constituting Britain as compassionate
(b) Constituting politicians as weak
(c) Constituting asylum seekers as bogus
(d) Constituting asylum seekers as a threat
(e) Constituting asylum seekers as a deviant social group

Constituting Britain as compassionate

Discursive studies (van Dijk, 1993, 1997b, Every and Augoustinos, 2007a) have noted how both the arguments for and against asylum seeking are

DOI: 10.1057/9781137415042.0006

almost invariably prefaced with a categorisation of the host country being a generous nation with a long tradition of hospitality towards foreigners who are in need of care and protection. In their analysis of pro- and anti-asylum seeker discourses in Australia, Every and Augoustinos (2008a) have noted how politicians on both sides of the debate employ the discourse of 'Australian generosity' towards outsiders. Similarly Jones' (cited in Every and Augoustinos, 2008a) analysis of the UK parliamentary debates also illustrates how emphasising British tolerance and humanitarianism was evident in both pro-asylum and anti-asylum debates. In the case of Australia, Every and Augoustinos (2008a, p. 570) argue that:

> 'generosity' is also used for different purposes in the Australian debates: to demonstrate that Australia's image would be tarnished by this legislation and thus should not be passed; but also to justify Australia as a 'generous nation' and the anti asylum seeker legislation as legitimate and necessary. This use of generosity by both sides demonstrates ... that it is not the 'content' of nationalism that identifies it as 'exclusive' or 'inclusive' *per se*, but the ways in which such constructions of the 'nation' are used. (emphasis in original)

The parliamentary debates that were selected for this chapter were in favour of tougher measures. They were also prefaced with similar claims of compassion as evidenced by Britain's long tradition of giving sanctuary to those fleeing political persecution. For example, in response to an emotive statement by Jeremy Corbyn, MP for Islington North, in which he gave a detailed and vivid description of the circumstances of Kurdish asylum seekers before questioning government policy of not providing sufficient resources and detaining Kurdish asylum seekers who had such an irrefutable long history of being persecuted in Turkey, the then Minister of State (Home of Office), Tim Renton, began his defence by stating that:

> I will begin by explaining the general context of the Government's policy towards people who claim asylum. As the hon. Member of Islington, North reminds us, the United Kingdom was one of the earliest signatories of the 1951 United Nations convention on refugees. We take our responsibilities very seriously, despite what is sometimes said by organisations such as Amnesty International ... No one who does my job can fail to be affected by the daily plight of people who are fleeing from persecution ... An application of asylum, I fully realise, as do Home Office officials, is important, sensitive and under our law must be properly and exhaustively considered ... No one is refused asylum until the full enquiries have been made ... If the interests of people

DOI: 10.1057/9781137415042.0006

> genuinely fleeing from persecution are to be safeguarded, it is vital that the system designed to protect them should not be exploited by people whose main motivation is economic migration. I want now to consider the particular circumstances of the recent influx of Turkish asylum claimants.... (*The Hansard*, 26 May 1989, Col: 1263)

Within this extract, Tim Renton takes particular care to construct the United Kingdom as a country that is compassionate and takes its international treaty obligations seriously. He also portrayed the nation as sensitive to the plight of asylum seekers, hence the careful consideration that is given to every asylum application as well as the strict adherence to the law. The effect of constructing the United Kingdom in this manner was that those who are refused asylum are then shown as not genuinely fleeing persecution but attempting to take advantage of the system and British generosity. This way the subsequent refusal of the claims for asylum and the subsequent banishment of claimants was presented as a reasonable, lawful and justified response to economic migrants masquerading as asylum seekers.

Four years earlier, Renton's predecessor had also in a similar fashion prefaced his defence of government policy with an emphasis on British hospitality. When asked how he could justify the first ever imposition of visa restrictions to a commonwealth country (Sri Lanka), the then Minister of State, David Waddington, began his response by stating that:

> People who are not refugees and do not belong here have not much to complain about if we say that in this small overcrowded country we have no room for them. But people who are not just coming here for a better life but are fleeing from persecution are entitled to special consideration, and we have always given such people that consideration. Our tradition of giving sanctuary to those fleeing from persecution goes back many years. Recently... we have given sanctuary to Poles, Iranians and citizens of many other countries who have made new lives here. I remind the House that the refugee statistics do not tell the whole story... In addition to those granted asylum on the basis of individual applications, we have admitted large numbers of refugees under specific programmes, most recently the 19,000 Vietnamese who do not appear in the refugee statistics as such. Many people who are not granted asylum are nevertheless allowed to stay exceptionally because of the conditions in the countries from which they have come. (House of Commons, 1985, 23 July 1985, Col: 971)

It is important to pay particular attention to the manner in which British generosity was foregrounded and worked up in the above account. The

DOI: 10.1057/9781137415042.0006

argument being advanced here was that in spite of having limited physical resources, Britain still fulfilled its moral obligations of looking after the needy and that it 'takes these responsibilities very seriously' (Tim Renton above). Britain's commitment to those fleeing persecution was presented as one that cannot be questioned as evidenced by the fact that Britain was one of the first signatories of United Nations convention on refugees. Almost invariably similar talk on asylum seekers advocating for tougher measures opens up with similar national rhetoric, which is replete with various forms of self-representation. These accounts were argumentatively constructed and oriented to attend to potential criticism that could be advanced in oppositional discourses, which could potentially undermine the notion of the British as a generous nation. As such, closely associated with this positive image of the British nation was the idea that in spite of its generous nature, Britain could not take in an infinite number of asylum seekers since it is only a small island. Van Dijk (1997b, p. 36) argues that these are 'the "national" correlates of what are known as face-keeping or impression management strategies in everyday interaction and dialogue'.

The rationale for providing such a positive self-image of the nation is primarily to provide a sharp contrast to the negative moral categorisation of asylum seekers, which then follows in which asylum seekers are portrayed as taking advantage and abusing British generosity. Such positive self-representation also has to be understood in the context of the politicians' quest to persuade the wider public that their actions were reasonable and justified. As such, it then appeared as though politicians were being forced by circumstances, in spite of the nation's natural benevolent disposition, to react harshly towards the apparent abuse of their generous nature by the 'evidently illegitimate' asylum seekers. This dualism has found expression in the well-known 'firm but fair' government policy statements. Although finding clearest expression in the 'firm but fair' approach, this repertoire evinces several features. Britain's compassionate outlook must be balanced against, 'the current labour market situation' (Phil Woolas, *Hansard*, 3 November 2009, Col: 38WS). Equally, Britain's 'long and proud history' of giving sanctuary to those with 'a well-founded fear of persecution' (David Hamilton, *Hansard*, 13 December 2005, Col: 1220) is deployed as a tactic for deflecting criticism for restrictive measures, in this case the imposition restrictions on economic migrants.

It is important to note that constituting Britain as compassionate as evidenced by its long tradition of receiving asylum seekers essentially

DOI: 10.1057/9781137415042.0006

served as an introduction to a real or mental *but*: Britain needs to remain realistic, Britain needs to be 'firm but fair', Britain needs to stop illegal migration, Britain needs to stop the 'bogus asylum seekers' who are abusing British generosity (van Dijk, 1997b). More recently, however, this strategy has morphed into the somewhat more muted call to 'integrate those who need our protection' (Home Office, 2006, p. 11). As illustrated in the last chapters, the balance of emphasis has shifted, quite obviously and decisively, towards the need for firmness, with particular emphasis being placed on the speedy processing of all claims and the prompt removal of unfounded claimants, evinced in the other strategies (Home Office, 2006, *passim*; John Reid, *Hansard*, 23 May 2006, Col: 76WS). The Coalition Government's Asylum Operating Manual launched in April 2013 is specifically designed to expedite decision-making. Nonetheless, as one government undersecretary of state put it, 'We give asylum to genuine asylum seekers. That is something that we can be proud of in this country' (Lord West of Spithead House of Lords, *Hansard*, 15 March 2010, Col: 449).

What is also clear from the foregoing is that the construction of asylum seekers within this interpretative repertoire is that the notion of nationhood was drawn on recurrently as a discursive resource. The responses to asylum seekers as the Other were formulated through invoking particular values of generosity that were associated with the nation state and the nation's character. According to Reicher and Hopkins (2001, p. 77) this was done 'in such a way as to invoke certain general values about the treatment of others – we are a tolerant nation; we are a nation which embraces otherness; we are above prejudice ....' The principal objective was to sidestep allegations of exclusionary tendencies and racism as the nation was projected as generous and welcoming of the Others, except that the Others were the ones that were taking advantage of *us* leading *us* into taking these stringent measures that we would not otherwise have contemplated.

Constituting the government as weak

In July 2013, David Cameron characterised asylum and immigration as 'out of control', 'a constant drain on public resources' and argued that this was the result of a decade of 'lax immigration policy' by the New Labour Government (Swinford, 2013). Within this interpretative repertoire, it was argued that asylum seekers were able to abuse British hospitality due to the weakness of the government as evidenced by its inability to

DOI: 10.1057/9781137415042.0006

manage effectively immigration and the ever-increasing number of 'bogus' asylum claims. According to Diane Abbot, MP for Hackney and Stoke Newington:

> The key to choking off unfounded asylum claims is to get rid of delays. The biggest single incentive for unscrupulous immigration advice telling people to stick in an unfounded claim is the delay. It has been common knowledge for a long time that it takes years to process an asylum claim... [The answer] lies instead in a fair, efficient and speedy system for both the initial decision, and crucially the appeal. (House of Commons, *The Hansard*, 24 April 2002a, vol. 384, Col: 409)

It was argued that because the existing government machinery could not deal effectively and decisively with the growing number of the 'bogus' asylum seekers, Britain was generally perceived as a 'soft touch'. The effect of portraying the government as impotent was that it evoked a sense of urgency to secure borders. David Lindlington, Aylesbury MP, expressed his concerns that:

> [the long] tradition of hospitality is undermined by large scale systematic abuse of our asylum law. The Government – by their negligence, complacency and incompetency – have undermined public confidence in the law and betrayed the interests of not only the British people, but of the refugees whom our law and international law have been developed to protect. The Government's record is dismal. (House of Commons, *The Hansard*, 24 April 2000a, vol. 348, Col: 435)

Therefore, playing out of this repertoire was evidently not only in the defence of government policy, which was supposed to be robust – '*choking off* unfounded asylum claims' (Diane Abbot, House of Commons, *Hansard*, 24 April 2002b, Col: 409) – crucially it was also oriented to opposition attacks on the 'large scale systematic abuse of our asylum law... [created by government] negligence, complacency and incompetency... [which has] betrayed the interests of not only the British people, but of the [genuine] refugees' (David Lidlington, House of Commons, *Hansard*, 24 April 2000c, Col: 435). In this context, the calls for tougher asylum legislation were justified not only on the basis of the need for an efficient administrative system, but also on the need to protect one of the putative core British values of providing a safe haven for the persecuted. The need for more stringent legislation was legitimated on the grounds of protecting the genuine refugee as well as restoring public confidence in the political system.

DOI: 10.1057/9781137415042.0006

Regardless of the putative willingness of politicians to expedite claims and repatriate bogus asylum seekers, a third tactic was to characterise the system as dysfunctional and its political masters as inept. Regarding lengthy delays in processing, the only two conclusions to be drawn were that they were either 'a sign of a failing and dysfunctional Department, or... the policy of this Government' (Stewart Jackson, House of Commons, *Hansard*, 26 October 2009, Col: 12). Again, this evoked a sense of urgency, which justified the calls for the enactment of a more restrictive asylum regime.

Constituting asylum seekers as bogus

Another strategy used in the selected media and parliamentary discourses was to depict asylum seekers as 'bogus'. Perhaps this is the clearest illustration of the logically precarious status of interpretive repertoires as asylum seekers were frequently depicted as 'bogus', even though the judgement assumed the outcome of the application process, which, by definition, had not been reached. The term 'bogus asylum seeker' first appeared in 1985 in response to the growing numbers of Tamil asylum seekers. However, the term bogus asylum seeker, and the bogus/genuine asylum seeker binary proved too much a temptation to resist and became buzzwords for both the media and the politicians. According to van Dijk (1997b), it became a convenient rhetorical strategy by which governments and politicians could justify the enactment of strict immigration controls. Lynn and Lea (2003) have illustrated how the existence of the 'bogus asylum seeker' and the 'economic refugee' within the United Kingdom has come to be regarded as common knowledge. This is in spite of the problems associated with these definitions and that strictly speaking there is nothing like an 'economic refugee'. Lynn and Lea (2003, p. 433) have stated that, 'the concept of "bogus asylum seeker" has become so "naturalised" within the UK, that from an argumentative viewpoint it is perhaps no longer necessary to defend the accusation that many asylum seekers are not fleeing from oppressive and hostile conditions in their home countries'.

However, the existence of this as common knowledge potentially has the drawback of allowing it to lose its persuasive appeal. In order to counter this possibility, the concept of 'bogus asylum seeker' is usually carefully linked to welfare provision to give it a powerful rhetorical appeal. Linguistic resources such as rhetorical contrast and extreme

DOI: 10.1057/9781137415042.0006

case formulation are also deployed alongside this concept. For instance, Lynn and Lea (2003, p. 433) illustrate the effectiveness of this strategy by referring to one of the letters to the editor they analyse in which the writer states that, 'Bogus ones are housed within weeks and the UK citizens, black and white, are left to rot in hostels ...' Evoking images of British citizens 'rotting' in hostels after preference is given to those asylum seekers, who clearly do not deserve it, is specifically meant to stir up emotions and foster a sense of resentment (Lynn and Lea, 2003). The concept of the 'xenos' assumes some importance in this depiction of asylum seekers as the Other. Asylum seekers are therefore constructed as outsiders/xenos who are perceived as an external threat that is threatening the values and ways of life of the British society. However, there is also a deliberate attempt to detach this depiction of asylum seekers from the conventional trappings of racism. In order to sidestep potential charges of being racist, there was an attempt by the writer cited above to show that he or she was equally concerned about the welfare of UK citizens, both black and white. In this context, the resentment towards the identified 'bogus' and therefore underserving asylum seeker is portrayed as an understandable and justified response. In fact, in parliament itself, some politicians have explained some of the public resentment towards asylum seekers in this way. For instance, the MP for Maidestone and The Wealde, Ann Widdecombe, referred to the resentment towards asylum seekers in Dover and emphasised that in order to attain good community relations '... a firm and fair immigration system is an absolute prerequisite' (*The Hansard*, 2 February 2000b, vol. 343, Col: 1048).

In order to give credibility to the distinction between the 'genuine' and the 'bogus' and the resentment towards the latter, arguments were placed within the bounds of what Billig (1988) characterises as 'reasonable' prejudice. This was often denoted by the ubiquitous disclaimer: 'I have nothing against asylum seekers but ...' Potter and Wetherell (1987) have defined disclaimers as pre-accounts which attempt to ward off anticipated negative attributions in advance of an act or statement. In this context, the use of disclaimers, as Hewitt and Stokes (1975, cited in Billig, 1988, p. 112) note, is a linguistic strategy called '... "credentialing": the speaker wishes to avoid being branded negatively and, in the case of prejudice, being someone who harbours unreasonable antipathies'. In addition, as van Dijk (cited in Billig, 1988) illustrates, the introduction of contrary themes, often with a connecting '*but*', served to provide a denial

DOI: 10.1057/9781137415042.0006

of prejudice and implied that the majority of the people were forced by factors outside their control to resent asylum seekers. According to Billig (1988, p. 113), the syntax of such statements 'tells its own psychological story: "we" are not resenting "them" of our own accord, but something, more often than not "them" themselves, are getting "us" to do the resenting'. This effectively shifts the blame away from the speaker squarely onto the 'bogus' asylum seeker. In relation to politicians, constructing asylum seekers as 'bogus' provides justification for the increasingly restrictive asylum legislative regime. Politicians therefore portrayed themselves as not prejudiced against immigrants but as being forced by circumstances beyond their control:

> In this sort of discourse, there is a denial of freedom. Things are happening – to make 'us' resent 'them', to make 'us' legislate against 'them' – which force 'us' a necessity, beneath which 'we' must necessarily bend. 'We' have to do things, even say things which we would not choose to do, feel and say if we were free from the yoke of necessary things. In this way the discourse employs a style which simultaneously deplores, denies and protects prejudice. (Billig, 1988, p. 114)

Thus, the need to act decisively was portrayed as pushing politicians into making difficult decisions that they would not ordinarily make. According to van Dijk (1997b, p. 36), 'this dualism is routinely expressed by the well-known "firm but fair" move: Pragmatic decision making requires that we are "firm" but at the same time remain "fair"'. It is also important to note that by emphasising the genuine/bogus binary, the argument that was advanced in both media and elite discourses was that they hold nothing against asylum seekers per se, so long as they were genuine. In so doing, both discourses avoided being positioned and labelled as overtly racist or xenophobic and at the same time the harsh treatment and exclusion from mainstream society of those deemed as 'bogus asylum seekers' became justified.

Constituting asylum seekers as a threat

Asylum seekers have also been constructed as a deviant social group that posed a 'credible' threat to western nations. The ideological purpose for constituting asylum seekers as a credible threat and the United Kingdom as a country that was under siege was to create a significant distance between 'us' and 'them'. The use of vivid lexicalisation established a wider polarisation between 'us' and 'them'. This was primarily achieved

DOI: 10.1057/9781137415042.0006

through the effective deployment of metaphors in the construction of asylum seekers in both media and parliamentary discourses.

A metaphor can be defined as a figure of speech in which a word or phrase that ordinarily designates one thing is used to designate another, thus making an implicit comparison. The utility of metaphors comes from their capacity to make use of 'one highly structured and clearly delineated concept to structure another' (Lakoff and Johnson, 1980, p. 61). This enables the understanding of pertinent issues as familiar terms are used to describe complex issues. These familiar terms which manifest themselves as vivid images have a 'normative' force that emanates from specific purposes and values as well as normative images that are deeply embedded within a specific cultural context (Schoen, 1993). The significance of metaphors as linguistic resources can be summed up as follows:

> [A] metaphor draws on the unconscious emotional association of words, the values of which are rooted in cultural knowledge. For this reason it potentially has a highly persuasive force because of its activation of both conscious and unconscious resources to influence our intellectual and emotional response, both directly – through describing and analysing political issues – and indirectly by influencing how we feel about things. (Charteris-Black, 2005, p. 30)

Therefore it is mainly through the direct connection that metaphors make with pre-conceived notions and images (as well as the emotions associated with such images) that metaphors can significantly influence how people think about asylum issues (Bleasdale, 2008). The fact that the meanings of the terms are familiar and shared by the targeted audience means that politicians can vividly and persuasively communicate their perceptions of asylum seekers.

Metaphors are an important tool in the legitimisation process, which is very crucial in political and media discourse. Conversely, metaphors are also important for the process of delegitimisation in these discourses. According to Chilton:

> Delegitimisation can manifest itself in acts of negative other-presentation, acts of blaming, scape-goating, marginalising, excluding, attacking the moral character of some individual or group, attacking the communicative cooperation of the other, attacking the rationality and sanity of the other. (Chilton, 2004, cited in Charteris-Black, 2005, p. 17)

As such, metaphors are used with both delegitimisation and legitimisation effects. In the case of asylum seekers metaphors are deployed as part of the attempts at negative Other representation.

DOI: 10.1057/9781137415042.0006

Recent studies (Finney and Vaughan, 2008, Klocker and Dunn, 2003, Finney and Robinson, 2008, Pickering, 2001) have analysed the use of metaphors by politicians and the media in their discussions of asylum seekers. These studies have illustrated how the use of metaphors depicts asylum seekers as 'different', the 'Other' and a 'threat' to the nation. The most common metaphors used in this context, for instance, were those that evoked images of 'floods', 'waves', 'tides', a 'deluge', massive 'flows' and 'swamp'. Finney and Vaughan (2008) have noted the widespread use of these metaphors in their comparative study of press representation of asylum seekers in Cardiff and Leeds. Buchanan (2001) also noted a heavy use of such metaphors in her analysis of media coverage of the closure of Sangatte refugee camp and the arrival of its former inhabitants in the United Kingdom. The use of similar metaphors can also be identified in parliamentary debates. For instance, in his speech on benefit reforms under the Immigration and Asylum Bill, the then Secretary for Social Security, Peter Lilley, stated that, it was 'crucial that the new procedures are not in turn overwhelmed by a rising tide of unfounded claims' (*The Hansard*, 11 January 1996, vol. 269, Col: 331). This deployment of metaphors of water and liquids in the depictions of asylum seekers served to provide a powerful and vivid image of a country that is overwhelmed and at risk of being submerged by waves of people. It also further reinforces the image of a small island with limited resources that is struggling to cope with the influx. Under such circumstances, the humanity of asylum seekers and their particular vulnerabilities were lost in the overwhelming sense that was generated by these volume terms (Bleasdale, 2008). More importantly, it also evoked a sense of urgency to stem the tide that threatened to submerge the nation. In this sense the metaphors functioned to make the stringent laws that are then advocated for and eventually enacted more acceptable as they were presented as a justified response to a credible threat to the nation.

Furthermore, asylum seekers were constituted through the meshing metaphors of war and criminality. Pickering (2001) captured the power behind war metaphors when she stated that:

> A war is only won or lost and there can be only one just side; only one force can maintain the high moral ground – the righteousness of one side's cause so great as to justify violence. The other is derided; impossible for them to assert the justness or legitimacy of their cause. Sides are therefore demarcated, boundaries and lines drawn ... In constructing a war, identities and individualities are irrelevant and excluded; there are simply sides – 'ours' and 'theirs'. (Pickering, 2001, p. 174)

DOI: 10.1057/9781137415042.0006

Thus, the effect of deploying war metaphors was to construct asylum seekers as the Other; the enemy. Buchanan (2001, p. 13) noted how military references and metaphors were employed throughout the coverage of the closure of Sangatte to describe the number, position and appearance of the refugees:

> when the camp closed to new arrivals at the beginning of November, the *Daily Express* described 'legions of young men'... looking like a 'rag tag army of conscripts' leaving the Sangatte camp. A few days later, an article in the *Express* warned of 'ranks of migrants' who were still 'massing at Calais', '... fuelling fears that French authorities are failing to stem the "flood of migrants"'. When refugees who had been turned away from Sangatte were invited to take shelter in a church by a priest, the negotiations by the French authorities to persuade them to leave were described as a 'siege' or 'stand-off' that was brought to an end by 'a dawn-raid'. The 'siege' ended with the refugees being surprised by police entering the church early one morning.

The effect of such a portrayal of asylum seekers was that it conveyed a vivid picture of a country that was under attack and needs to act decisively to deal decisively with the threat that asylum seekers posed.

Newspaper headlines were also sometimes formulated using such war metaphors. For instance, the *Daily Mail* on 26 November 2002 had a story entitled *Losing the war on asylum crime*. The overall effect of these strategies was that asylum seekers were portrayed as a credible threat not only to national security but also to the hegemonic ways of life of those falling within the remit of the *us* in-group. Instead of being 'presented as people who are trying to escape threat, they are, in most cases, presented as the threat' (Bailey and Harindranath, 2005, p. 283).

Furthermore, constructions of asylum seekers in the press also deployed metaphors that emphasised the illegality of asylum seekers and their 'disposition' to crime. For instance, an article in the *Daily Mail* linked asylum seekers to criminal activities by cataloguing 44 serious crimes committed which included rape, murder, sexual assault and fraud (Williams, 1998). The article went on to explain how the taxpayer by default was made to directly and indirectly fund these criminal activities. One of the *Daily Star's* headings on 22 May 2002 claimed that *Asylum Seekers: 9 out of 10 are conmen*. The use of war metaphors and references to criminality had the effect of constructing the relationship between asylum seekers on the one hand, and the government and the rest of the citizens on the other, as conflicted with the potential of a violent showdown. According to McLaughlin (1999), such use of war

DOI: 10.1057/9781137415042.0006

metaphors gives rise to discourses of war which in turn promote the need to repel whatever is hostile and threatening. Pickering (2001, p. 173) has also argued that 'such representations contribute to the validation and invocation of repressive state responses'. Therefore, enmeshing immigration and criminal discourses with discourses about tactics of war further legitimised the harsh treatment of asylum seekers and their eventual exclusion. This is justified on the grounds of the need to protect the nation state as exemplified by the moves towards making borders even more secure. Thus, the use of war metaphors further narrowed the discourse on asylum seekers into one about nationhood. Significantly, the discursive repertoire of war rendered acceptable and reinforced a sense of normalcy about the government's responses, which would otherwise have been questionable and potentially met opposition.

Constituting asylum seekers as a deviant social group

In addition to being constituted as a threat, asylum seekers were also categorised as a deviant social group. Asylum seekers were portrayed as 'not like us'. This was usually achieved through the use of ethnic markers such as identifying asylum seekers in terms of being 'Somalis', 'Afghans', 'Rwandans', 'Iraqis', etc. A significant proportion of asylum seekers in the United Kingdom are Muslim. With reference to Australia, Karim (1997) has argued that people are socialised in a manner that implicitly and explicitly encourage them to identify Muslims as the 'Other', primitive, uncivilised, terrorist, the bad guy, female oppressor and innately prone to violence among other things. In the post-9/11 context, such ethnic marking consolidates the construction of asylum seekers as the enemy 'Other' (Rashid, 2007). The connection with terrorism also connects strategies (c) and (d), where asylum seeking is presented as providing a means of assistance for terrorists, or for deferring justice. 'How many of those arrested on suspicion of terrorist offences since 11 September 2001 have at some point claimed asylum?' (Patrick Mercer, *Hansard*, 6 May 2009, Col: 267W; question for Home Secretary; see also Frank Field, *Hansard*, 6 May 2009, Col: 267W).

In addition metaphors of disease were also employed in constituting asylum seekers, thereby underlining the need to exclude and expel them. The implicit 'threat' of sickness that asylum seekers supposedly posed was not hard to discern in political and media discourses. The interpretative frameworks of race and the integrity of the nation state provided a broader and additional context for the understanding of the ways in

DOI: 10.1057/9781137415042.0006

which the metaphors of disease were deployed in the construction of asylum seekers as a deviant social group and how this then underscored the imperative to exclude and eventually expel them. With reference to Australia, Pickering (2001) has illustrated how the implicit 'threat' of disease that asylum seekers pose is not far below the surface of dominant anti-asylum discourses. She evidenced this by pointing at the requirement for asylum seekers to undergo 'health screenings' and 'medical checks' even though 'what people are being screening and checked for was largely absent' (Pickering, 2001, p. 181). Similar perceptions are also discernible in the United Kingdom. For instance, in 2003, the Institute for Public Policy Research warned the government against the introduction of compulsory health screening, which was being considered as part of the Asylum Bill (Leudar et al., 2008).

The deployment of discourses of disease served a double purpose here. On the one hand, in a literal sense, asylum seekers were constructed as a burden to the NHS and also posing a significant danger to public health. Asylum seekers' deviancy as a social group was constructed in terms of how as a 'diseased' group they 'exploited' the national health system at the expense of the nation. This perspective has indeed been influential in informing some anti-asylum media discourses in the United Kingdom. For instance, in a letter to the editor in the *South Wales Echo* one reader expressed her concerns as follows:

> it has been decided that illegal immigrants, failed asylum seekers and refugees who are waiting appeal to remain in the UK will now be given free healthcare at a new practice based at Cardiff Royal Infirmary. This will involve screening service for communicable diseases plus the provision of interpreters, putting an even greater burden on the NHS, which is breaking at the seams. While the indigenous population are expected to work up to 70 years of age, are left waiting on trolleys, operations are cancelled and people are having to sell their homes to pay for life saving drugs, certain immigrants who have no right to remain in the UK are to receive specialist treatment courtesy of the barmy army brigade. (South Wales Echo, 3 September 2008, p. 18)

Another reader also wrote to *The Express* complaining that 'Britain can no longer be the NHS of the world' and also expressed his bitterness over what he regarded as preferential treatment to asylum seekers. He stated that 'British people of all races have paid their taxes for years but are being denied treatment by outsiders who have not contributed' (*The Express*, 29 July 2004, p. 57). Another writer also expressed his astonishment that 'asylum seekers and immigrants with HIV are allowed into this

DOI: 10.1057/9781137415042.0006

country. So why are our politicians and government not protecting our people?' (*The Express*, 13 August 2002, p. 30). In these extracts, asylum seekers were constructed as a deviant social group that threatened the health of the nation, the very survival of the NHS, and were also resented because they were queue jumpers.

On the other hand the formulation of asylum seekers as a diseased social group slips seamlessly into the metaphorical construction of asylum seekers as a pathological presence in the host social body. Pickering (2001, p. 182) has aptly summed up the impact of constructing 'asylum seekers' through metaphors of disease:

> Disease sees asylum seekers constructed not only as problems, but also as deadly problems. In becoming linked to the transmission of disease an analogy is created: asylum seekers threaten the life of the host society – a society that is repeatedly presented as the healthy and the robust and the asylum seeker as the foreigner (pest), the polluted enemy that potentially compromises the health and endangers the wellbeing of the nation. (Pickering, 2001, p. 182)

Thus, the deployment of metaphor of diseases in the construction of asylum seekers has the overall effect of depicting asylum seekers as the undesirable Others whose expulsion is necessary for the health of the nation.

Dissenting voices

Although asylum seekers tended to be predominantly constructed in negative terms as illustrated, discursive studies (Every and Augoustinos, 2008a, Lynn and Lea, 2003, Finney and Robinson, 2008, Every and Augoustinos, 2008a, 2008b) have also identified dissenting voices that offer counter discourses. An appreciation of this will enable social work to reject the inevitability and sense of 'expectation' that the kinds of treatment and services asylum seekers receive are necessarily part of being an asylum seeker. Such an understanding may help enable social work to unambiguously assume a subject position in ensuing discourses and reorient social work towards the social justice agenda and to see through media and political representation of asylum seekers.

In their study of accounts that oppose the dominant discourses on asylum seeking in Australia, Every and Augoustinos (2008b, p. 653)

DOI: 10.1057/9781137415042.0006

have identified seven strategies that are commonly employed in counter discourses:

(1) The dissenting voices foreground the similarities between asylum seekers and Westerners/Australians. They argue that in similar circumstances 'we would do the same', that is, flee our country and seek protection from other safer countries. This way they attempt to challenge the 'us' and 'them' dichotomy by re-categorising asylum seekers as *just like any one of us*.
(2) They also attempt to draw comparisons between present-day asylum seekers and previous ones such as Jews during and after the Second World War.
(3) They also draw on logic in their attempts to undermine the dominant anti-asylum discourses.
(4) They also employ metaphors and analogies in their descriptions of asylum seekers.
(5) They also try to undermine the validity of the dominant discourses by drawing on those discourses that potentially may resonate with an Australian/Western audience such as the need for families to stay together rather than attributing asylum seeking to factors such as soft policies, economic migration and personal choice.
(6) They also further argue that asylum seekers are *victims of circumstances* emphasising that they had no choice but to leave their countries of origin and as such they deserve better treatment. In this way the counter discourses are able to invoke a moral obligation towards asylum seekers.
(7) They also re-lexicalise commonly used and abused terms like 'persecution' with more emotive terms to achieve greater impact.

These strategies are quite relevant as they also feature quite frequently in social work professionals' discourses that are discussed in the next chapters. Within the United Kingdom, counter discourses seek to construct asylum seekers differently from the ways they are portrayed in dominant discourses. For instance, Lynn and Lea (2003, p. 442) refer to a letter to the editor that appeared in *The Guardian*, on 3 September 2001, in which the asylum seekers were described as follows:

> we should remember that asylum seekers are locked up to have their claims processed at the Oakington immigration centre... This is a reception centre in name only: even the Home Office has admitted that it is, in fact and in

> law, a detention centre. People detained there include children, pregnant women, the elderly, the ill and survivors of torture.... (cited in Lynn and Lea, 2003, p. 422)

Such counter discourses seek to expose truth behind institutional rhetoric. According to Lynn and Lea (2003) the above account cuts across the 'official' language in its attempts to deconstruct the 'reception centre' and show what it really is – a jail. This way, the inhumane treatment of locking up innocent people just to allow bureaucratic and administrative procedures to be completed is brought to the fore. The 'fact' that the Home Office conceded that the 'reception centre' is 'in fact and in law, a detention centre' makes the argument even more persuasive and convincing. As Edwards and Potter (1992) state, facts are quite difficult to refute. Referring to the law to define the 'reception centres' as 'detention centres' is an effective strategy that also leaves very little room for a counter argument. According to Leishman confession is the 'king of evidence' (cited in Lynn and Lea, 2003, p. 443).

Furthermore, the counter discourse deployed vivid rhetorical contrasts in reconstituting of asylum seekers. For instance, in the case of the letter cited above, rhetorical contrast was presented in the form of the long list of those detained who are vulnerable which includes children, pregnant women, the elderly, the ill and survivors of war and torture who were detained not because they had committed any crime but for administrative and political conveniences. Such linguistic strategies evoked images of the concentration camps and a totalitarian state. According to Lynn and Lea (2003, p. 443),

> Lists and contrasts are powerful rhetorical devices, and a more vulnerable and disempowered group of individuals as that [re]presented here, would be difficult to find. There appears to be little that is bogus or threatening about them; the image they conjure is contrary to that perpetuated by common knowledge. Here then, is a counter-discourse, which emphasizes how asylum-seekers are incarcerated, regimented and controlled in the most oppressive fashion.

Therefore, the counter discourses attempt to reclaim the humanity and individualism that asylum seekers are stripped of by the dominant discourses. The ability of the counter discourses to reclaim the humanity and individuality of asylum seekers demonstrates the 'stark face of a political ideology at work, or as Foucault described it: "power in practice"' (cited in Lynn and Lea, 2003, p. 443). The dissenting voices reject

DOI: 10.1057/9781137415042.0006

the inevitability and sense of 'expectation' that the kind of treatment asylum seekers receive is necessarily part of being an asylum seeker. As a result of this, the counter discourses then go further and attribute to government the same qualities of artifice and duplicity that were the preserve of asylum seeker in the dominant discourses.

However, the dissenting voices were very much in the minority. As such, their impact was largely minimal. Nonetheless, the challenge they posed to the dominant discourses on asylum seekers was significant enough to warrant counter arguments from the dominant discourses which sought to further marginalise them (Lynn and Lea, 2003). For example, during the debate on the proposed social benefits for asylum seekers under the 1996 Asylum and Immigration Bill, Liz Lynn, MP for Rochdale, challenged government policy stating

> Is it right that persecuted people should now become homeless and destitute? A number of genuine asylum seekers who apply in-country have been tortured. Should not they be treated with compassion, and not kicked in the teeth? (House of Commons, *The Hansard*, 11 January 1996, Col: 337)

The response from Peter Lilley was that, 'The hon. Lady's contribution was characteristically emotive and irresponsible. While this is an emotional subject, it is not a matter on which one should be emotive, which means stirring up unjustified emotion' (House of Commons, *The Hansard*, 11 January 1996, Col: 337). Such a rebuttal is indeed characteristic and a ubiquitous response to similar challenges to the dominant anti-asylum discourses. It served to reiterate the need for the politicians to be pragmatic about the job at hand.

Another characteristic response of dominant discourses to such dissenting voices was to differentiate the self even further. This was achieved by identifying what Lynn and Lea (2003) have characterised as the 'enemy within' – the white liberals. Since categories are for talking (Edwards and Potter, 1992), the white liberals were identified as a distinct entity that became the subject of attack in the dominant discourses' retort. The white liberals were singled out as pro-asylum and espousing humanitarian and *welfarist* discourses, but in an allegorical and sardonic way (Lynn and Lea, 2003). Within such a cynical context, the very nature of those categorised as 'white liberals' was then regarded with derision. For instance the credibility of 'white liberals' was undermined by characterisations of being rich, patronising and often saying one thing but doing exactly the opposite. In another letter analysed by Lynn and Lea

DOI: 10.1057/9781137415042.0006

(2003), white liberals were accused of taking humanitarianism too far yet taking care to ensure that they themselves were not disadvantaged economically in the process. This way 'implicitly, the credibility of counter-discourses is challenged as suspect, it is as untrustworthy as the people who promote it' (Lynn and Lea, 2003, p. 446).

Discussion

It is clear from the foregoing that talk about asylum seekers is not inherently xenoracist. The extent to which such talk can be judged as xenoracist was largely dependent on the interplay of discourses. Xenoracism is not inherent in asylum seeking discourses but rather it is an *effect* of using specific discursive and rhetorical devices. The deployment of such rhetorical devices achieves specific acts such as blaming and positioning asylum seekers as beyond the moral order. Within the anti-asylum seeking discourses there was an absence of blatant expressions of prejudice which explicitly draw on biological categorisations and skin colour, yet the overall effect of the discourse was racist and exclusionary. At times, the term asylum seeker served as the marker of legitimacy and tolerance in discourses aimed at limiting and removing the 'bogus' economic migrant. When it was asylum itself that was being considered, asylum seeking was portrayed as a 'vulnerability' that opened up the system to abuse. What was even more striking was that this shift from 'testament to British compassion', to the spectre of the asylum seeker as menace was so effortless, and that they were not at all exclusive as strategies. It is in this sense that it can be argued that these various interpretative repertoires bear all the hallmarks of the xenoracism in terms of the subtlety of the exclusionary tendencies. Within the anti-asylum seeking discourses such tendencies can be characterised more accurately as xenoracist. This is because they contributed to an overall discursive strategy that was geared towards the exclusion of asylum seekers from mainstream British society.

An analytic focus of this chapter has been the analysis of the media and parliamentary discourses in their rhetorical (i.e., dialogic and argumentative) contexts. According to Billig (1988) one of the key features of accounts that are worked up in an argumentative context, such as the accounts presented in this chapter on the emotive topic of asylum seekers, is that they have contrasting versions. Due to the fact that the

DOI: 10.1057/9781137415042.0006

various constructs of asylum seekers were heavily contested, each of these discourses was influenced and indeed shaped by the other. This is why Billig (1988) has argued that discourses are constructed argumentatively. As such, both sides tended to orient their arguments in such a way as to undermine the other. According to Every and Auguostinos (2008b, p. 652), the various discourses 'not only counter arguments voiced within the parliament itself, but also recurring in the media and in the broader public opinion voiced on talkback radio, letters to the editor and opinion polls'. As such the discourses were argumentatively organised to explicitly or implicitly undermine or rebut alternative or opposing accounts. Therefore, these types of discourses can best be understood if they are analysed in their respective argumentative contexts which take into consideration those accounts that they are oriented to as attempted in this chapter. This has enabled a better understanding of the ways in which certain representations of asylum seekers are articulated.

Quite significantly, this chapter also serves to provide an important background for understanding some of the arguments that social work professionals' discourses were oriented to. As will be noted in the next chapter, social work professionals drew on some of the discourses, which are discussed in this chapter, and also responded to some of the arguments that were advanced in this chapter. By drawing on these macro- or general and historically positioned discourses, such as the media and parliamentary discourses presented here, as a discursive resource, an analytic link is established with the local or micro-discourses associated with the mundane day-to-day social work practice with asylum seekers. As such, in order to fully understand the local discourses associated with social work practice with asylum seekers, it is imperative that macro- or general and historically positioned discourses such as the ones presented in this chapter are also taken into full account as they have a significant influence on the local discourses themselves. The macro-discourses serve as a discursive resource and additional framework that has a significant influence on the orientation of the local discourses. As such, the meaning making process that social work professionals engaged in their attempts to make sense of their mundane day-to-day tasks was embedded in a much wider context. Although the focus of this book is on how social workers made sense of their work with this service user group, it was also apparent that the local meanings that were produced were situated and contextualised in larger and wider discourses such as the ones that have been presented in this chapter.

DOI: 10.1057/9781137415042.0006

However, even the implicit dichotomy of 'pro-' and 'anti-' that is attempted in this chapter only serves to provide a useful heuristic because the nature of the prevailing discourses is never static. In line with the rule of tactical polyvalence, discourse should be regarded as characterised by constantly shifting power relations:

> we must not imagine a world of discourse divided between accepted discourse and excluded discourse, or between the dominant and dominated one; but as a multiplicity of discursive elements that come into play in various strategies. (Foucault, 1998, p. 100)

As such, anti-asylum and pro-asylum discourses cannot be analysed in isolation as they are intertwined, with the same elements sometimes pressed into service of either side. Given that the various constructs of asylum seekers are heavily contested, each of these discourses is influenced and indeed shaped by the other. It is important to emphasise that although the distinction made in this book between pro-asylum seeking discourses and anti-asylum seeking discourses is a useful heuristic, the very act of making such a distinction reveals their interconnectedness.

DOI: 10.1057/9781137415042.0006

3
Asylum Seekers in Social Work Discourse

Abstract: *This chapter reviews the emerging social work literature on asylum seekers. It is noted that within this emerging literature there is a paucity of research that focuses on the ways in which asylum seekers are constructed as a social group and the implications for practice. The chapter locates the importance of language and discourse within social work. In so doing the chapter provides further justification for the book's focus on professional social work discourses of asylum seekers.*

Masocha, Shepard. *Asylum Seekers, Social Work and Racism*. Basingstoke: Palgrave Macmillan, 2015.
DOI: 10.1057/9781137415042.0007.

DOI: 10.1057/9781137415042.0007

Introduction

Within the United Kingdom, it can be noted that there is an emerging body of research that has resulted in the development of a social work discourse relating to asylum seeking service users. This body of social work literature is relatively new and is largely sympathetic to the plight of asylum seekers. This emerging discourse asserts that working with asylum seekers is a legitimate area of practice and highlights social work's obligations towards asylum seekers by exploring some of the salient aspects of the oppression that is suffered by this service user group. As such, it focuses on the negative impacts of current social policies on welfare provisions for asylum seeking service users. For instance, Humphries (2004a) and Hayes (2002, 2004) demonstrate the impact of immigration controls on social policies relating to welfare provisions for asylum seekers. Briskman and Cemlyn (2005) discuss the role of social work with asylum seekers in the light of the restrictive and repressive UK social policies relating to this service user group. The discourse uses this as evidence to underscore the fact that working with asylum seekers is a neglected area of social work practice. This is done through highlighting the fact that in spite of the apparent need for social justice for this group of service users, this is still not regarded within social work as an important area of practice deserving special attention. For instance, Sales (2002a) discusses how asylum seekers generally are perceived as a social group that exists on the fringes of British society and are regarded as undeserving of existing mainstream welfare supports. Similar research by Harris (2003), Masocha (2008) and Masocha and Simpson (2011b) also depict asylum seekers as an out-group. As such, this emerging body of knowledge draws attention to the various ways in which asylum seekers as recipients of social work services are a marginalised group and how this runs counter to what social work as a profession stands for as well as its underpinning principles and core values. For instance, Jones (1998, 2001) has suggested that social workers are ignorant of immigration law and its primacy over children's legislation in relation to unaccompanied asylum seeking children. Article 22 of the United Nations Convention on the Rights of the Child (1989), which came into effect in the United Kingdom in 1992, stipulates that unaccompanied asylum seeking children should be "accorded the same protection as any other child permanently or temporarily deprived of his or her family environment for any reason".

DOI: 10.1057/9781137415042.0007

This lack of understanding has resulted in asylum seeking service users not receiving the best service. Jordan and Jordan (2000) have argued that many social workers still do not accept that the social problems faced by asylum seekers are part of their core business. Such views and attitudes are likely to negatively impact on practice and result in asylum seekers' issues not being prioritised within social workers' workloads. Such views can also have a negative impact on social workers' ability to effectively advocate for this service user group.

An important feature of this emerging discourse is the tension that arises between the dictates of current social policies relating to welfare provisions for asylum seekers and what social work as a profession stands for. Here the research literature focuses on the ethical dilemmas that emerge for social work as a result of this tension. For instance, Humphries (2004a, 2004b) and Collett (2004) have separately demonstrated how social workers are increasingly assuming the roles of constriction and gatekeepers of the welfare system and are becoming an extension of the immigration service. They argue that in so doing, social workers are unwittingly being implicated in racist practices and in the process negating social work's commitment to social justice for less privileged social groups like asylum seekers. According to Humphries (2004b), it is in the area of anti-oppressive practice that the rhetoric of post-Thatcherite politics and the values to which social work aspires converge. The following themes have assumed increasing importance in post-Thatcherite politics: choice, citizenship, autonomy and social justice. These themes seem quite compatible with social work values and its commitment to strive for social justice for the less privileged in society. The problem is, while seemingly liberal policies are espoused, there is also a consensus that has emerged in which policies that systematically degrade and disenfranchise asylum seekers are called for and enacted, and social workers are obliged to implement them as demonstrated in Chapter 1. However, more recent research provides a much more positive view of social work practice. For instance, Kohli (2007, p. xxi) rejects the negative views of social work practice with unaccompanied asylum seeking children that tend to be portrayed by advocacy-based studies:

> In a broad sense therefore the book rejects the common conceptions of unaccompanied children as victims. It displays their vulnerabilities alongside their strengths within the varied attempts they make to salvage their lives with the UK. Similarly, it seeks to establish social-work practice as varied and rich in meaning for practitioners themselves, not as a deficit-laden activity.

DOI: 10.1057/9781137415042.0007

Similarly, Fell and Fell (2013) demonstrate the existence of an innovative and supportive practice with asylum seekers in the voluntary sector in spite of the largely constrictive environment. Masocha (2014) illustrates how social work practitioners encounter ethical dilemmas in their practice with asylum seekers and are clear in their understanding that these dilemmas originate from organisation structures, legislative frameworks and policy provisions which regulate their practice. He also noted how these impediments were "distinctly at odds with the practitioners' personal [and professional] beliefs of how asylum seekers should be treated as well as social work's orientation towards the social justice agenda" (Masocha, 2014, pp. 1632–1633). Thus a more positive view of social work practice is also offered within this emerging social work discourse on asylum seekers.

Significantly, this emerging discourse on asylum seekers also seeks to achieve a better and more comprehensive understanding of some of the needs of this service user group and how best these needs can be met by a more enhanced social work role. In this respect, this emerging body of knowledge tends to concentrate particularly on two prominent areas of practice: mental health social work with asylum seekers and social work with unaccompanied asylum seeking children. For instance, Chantler (2011) argues that mental health social work is in need of urgent attention as current practice falls below expectations. Chantler cautions against a heavy dependence on post-traumatic stress disorder as a diagnosis that can help in understanding comprehensively mental health distress experienced by asylum seeking service users and suggests a social model that pays particular attention to the wider circumstances of those with insecure or unresolved immigration status. Along the same lines, Masocha and Simpson (2011a) have also provided an account of the causes of mental difficulties that are experienced by asylum seeking service users. They provide a critique of the research literature that currently informs mental health social work with asylum seekers. They demonstrate that the existing literature mainly comes from psychiatric studies, which are informed by a medical model. They argue that it is untenable to try and understand the mental health difficulties that are experienced by asylum seekers solely based on aetiological accounts that draw heavily from psychiatry and are based on biological causation. They present a more comprehensive model for understanding the mental health difficulties experienced by asylum seekers. The model considers both biological causation and a social perspective and as such locates the mental health difficulties experienced by asylum seekers in a much wider

DOI: 10.1057/9781137415042.0007

economic and socio-political context. The model suggests areas for the development of mental health social work practice with asylum seekers as it maps out some of the areas where practice enhancements can be realised. Furthermore, Kohli and Mather (2003) have also suggested ways that social workers can work effectively to promote the psychosocial well-being of unaccompanied asylum seeking children in the United Kingdom. They suggest an approach that takes into cognisance both the vulnerabilities and the resilience of asylum seeking children.

In a much similar way, the existing research on social work with unaccompanied asylum seeking children is also relatively recent and very much a response to a lack of knowledge on this service user group. Research studies on social work practice with unaccompanied asylum seeking children have resulted in a discourse that centres on policies and practice issues of concern. The research has also drawn attention to not only the needs of this service user group but has also highlighted some of the complexities and practice dilemmas that social workers have to contend with. Munoz (2000) discusses how three London boroughs respond to the needs of 16- and 17-year-old unaccompanied asylum seeking minors and highlights the inadequate levels of support. Munoz's (2000) study highlights practice issues of concern which include the increased use of emergency accommodation which results in landlords becoming de facto guardians, the increased use of overcrowded hostels giving rise to child protection concerns and the lack of a robust monitoring and support system which exposes the unaccompanied asylum seeking minors to risks of significant harm. Extensive research by Professor Ravi Kohli has made a significant contribution to the understanding of the needs of unaccompanied asylum seeking children as well as suggesting ways to enhance practice. This has helped to foster an understanding of:

- The meaning of silence and secrets that unaccompanied asylum seeking children may present with and how best social workers can work in a therapeutic way in such a context (Kohli, 2006b, 2009,).
- The meaning of food to unaccompanied asylum seeking children looked after by local authorities. The study explores the powerful relationship that can exist between food and finding a safe haven, and even becoming an integral part of the foster family as well as the wider host community (Kohli et al., 2010).
- The meaning of safety, belonging and success for unaccompanied asylum seeking children (Kohli, 2011).

DOI: 10.1057/9781137415042.0007

Furthermore, studies by Dixon and Wade (2007) and Wade (2011) explore the challenges that pathway planning presents for social workers working with unaccompanied asylum seeking children. These studies underscore the importance of preparation and planning for transition to adulthood when working with adolescent unaccompanied asylum seeking service users. It is argued that these should be priority tasks and at the forefront of the minds of their care-givers and social workers. The studies also review the existing research studies on social work responses in England, with a particular focus on what is known about best practice in terms of preparing adolescents for adult life and on the challenges that the intersection between social work and the asylum determination process poses for pathway planning. Kohli (2007) offers a detailed view of social work practice with unaccompanied asylum seeking children who are looked after by local authorities under section 20 of the Children Act 1989. The study underscores the need to view unaccompanied asylum seeking minors as children first and foremost as well as the need to view social work as an important therapeutic activity that can help this group of service users to establish some roots and stability in their host country.

However, it is notable that one important consideration that is absent within the reviewed social work literature and this emerging discourse is a critical appraisal of how asylum seekers as a service user group are constructed and its implications for practice. The next two chapters seek to fill in this gap by specifically analysing how asylum seeking service users are constructed in social work professionals' discourses.

Framing professional discourse

An appreciation of professional discourses will not only result in an understanding of social work processes but will also enhance the understanding of their historical roots. The important role that is played by discourse within the construction and reconstruction of the social work profession is undeniable. Written texts, spoken discourse and various forms of non-verbal communication have played crucial roles in the historical construction of social work practices, and they continue to contribute to the reproduction and reshaping of these practices (Hall et al., 1997). In spite of the pivotal role that discourse occupies within social work, the interest in the understanding of the dynamics of professional discourse is quite new and largely undeveloped among

DOI: 10.1057/9781137415042.0007

social work researchers and practitioners. This is in spite of the fact that discourse is one of the central vehicles of social work. Social workers spend a lot of their time talking with service users and other professionals. Invariably these discursive acts are closely intertwined with, preceded and followed by writing practices leading to the production of reports, emails, contracts, etc. According to Hall et al. (2006, p. 10):

> These mundane activities of daily professional life are not merely insignificant processes which merely operationalise, facilitate or frustrate evidence-based practice or critical reflection. On the contrary, the objectives of social work can only be realised through such mundane activities and these practices do not just have an influence on social work, *they constitute it*, they bring it into existence. (emphasis added)

As such, discourse plays a central part in social work, hence the need for an in-depth study of its role in the construction of asylum seeking service users within practice.

In spite of the centrality of language in social work activities, within the United Kingdom research, which pays particular attention to how language is used to enact socio-cultural perspectives and social identities, has largely been outside the realm of social work research. However, there are few notable exceptions to this. Fook (2002) and Rojek (1988) have analysed how a knowledge of language use and discourse might be translated into useful practice strategies for practitioners. White (2003) and Hall et al. (2006) analyse language use to understand the various ways social workers, their clients and other professionals categorise and manage the problems of social work in ways which are rendered understandable, accountable and which justify professional intervention. Hall (1997) discusses the everyday activities of social workers as performances of storytelling and persuasion. Parton and O'Byrne (2000) analyse the relevance of social theories associated with postmodernism, social constructionism and narrative approaches to social work.

The next two chapters also use discursive psychology as a methodology for analysing the various ways in which asylum seekers were constructed through identifying the various interpretative repertoires that were employed in the construction of asylum seekers by social work professionals within their work settings. Sarangi and Roberts (1999, p. 1) define the work place as a 'social institution where resources are produced and regulated, problems are solved, identities are played out and professional knowledge is constituted'. In light of this definition, within a social

DOI: 10.1057/9781137415042.0007

work context the professionals can be accepted as constantly defining and redefining their respective discourses and perceptions of asylum seeking service users as part of an ongoing process which in turn gives rise to the social production of meaning. As such, the workplace can be regarded as a place where local discourses are produced. Hall et al. (1997, p. 281) characterise the everyday interactions on the work floor as 'local discourses' in their own right given that they provide social work stories that 'are constrained by the kind of discourse practices that are available to them and are currently used'. Therefore, in this book, the local interactions within social work practice with asylum seekers will be viewed as local discourses. The following chapters explore how such local meanings were constructed with specific reference to social work professionals' perceptions and constructions of asylum seeking service users.

The previous chapter focused on how asylum seekers were constructed at the meso and macro levels in the media and parliamentary discourses. However, it is not suggested here that there is a top-down determination of practice. Rather this book represents an attempt to examine the production of linguistic resources at that level and to discuss how it functions to constitute subjectivities in the welfare apparatus. Again, however, it is also not suggested that it is the only source of discursive resources or that social work practice is affected solely, or even primarily, at this level. The everyday interactions within social work practice with asylum seekers are influenced by general discourses in society. Social work practices relate to what Fairclough refers (1995) to as 'general' or 'societal' discourses that are provided by the society at large in addition to the professional discourses. Pithouse and Atkinson (1988, pp. 187–188) have suggested that 'within the social work setting, the actors' mundane theory is a contradictory amalgam of formal social work concepts, practice wisdom and the workers' understanding drawn from their participation in the wider culture beyond the formal work setting'. This means that social workers have other frames of reference outside their professional discourses, which they draw upon. Indeed, as will be noted in Chapters 4 and 5, social workers draw on politicians and media representations in their attempts to make sense of asylum seekers and related practice issues.

Therefore, whereas Chapters 1 and 2 focused on the macro and historically positioned asylum seeking discourses, the subsequent chapters are devoted to exploring how at a more local level social work professionals

DOI: 10.1057/9781137415042.0007

construct asylum seeking service users and pay particular attention to the linguistic resources that they deploy. Given how asylum seekers are constructed in the media and parliamentary discourses and how social workers are not immune to these discourses, the previous two chapters also serve to provide a wider background for understanding some of the resources that the social workers interviewed drew upon in their attempts to make sense of this service user group. In this sense, the media and parliamentary discourses are accepted as providing some of the additional frames of reference that were available to social work professionals in addition to the professional repertoires, institutional resources and frameworks. The subsequent chapters analyse some of the interview data from 25 local authority social work professionals who at the time when the study was carried out were working with asylum seekers in various capacities. The chapters identify and analyse a number of interpretative repertoires that were discernible in the social work professionals' narratives.

DOI: 10.1057/9781137415042.0007

4
Countering Hegemonic Narratives

Abstract: *This chapter explores how social work professionals counter the hegemonic anti-asylum seeking discourse. It pays particular attention to the specific subject positions that were taken up by practitioners in relation to the dominant anti-asylum seeking discourses. The data analysis revealed a number of interpretative repertoires, which were utilised by social workers to provide counter narratives to the dominant ways of portraying asylum seekers. Within these interpretative repertoires various linguistic strategies were employed by social workers interviewed in their attempts to deconstruct and reconstitute asylum seeking service users. The chapter demonstrates that by engaging in these discursive acts, social workers were able to portray positively not only asylum seekers but also the profession as a whole. The social workers' dissenting voices were able to reject the inevitability and sense of 'expectation' that the kind of treatment asylum seekers receive was necessarily part of being an asylum seeker. By engaging in such discursive acts and taking up specific subject positions, the practitioners were able to foreground the humanity and individuality of asylum seekers. From an advocacy standpoint, the chapter underscores the importance of social work being actively involved in discursive acts that constitute those particular groups' identities.*

Masocha, Shepard. *Asylum Seekers, Social Work and Racism*. Basingstoke: Palgrave Macmillan, 2015.
DOI: 10.1057/9781137415042.0008.

Introduction

Limited research has been carried out from a discursive perspective to explore the specific subject positions that were taken up by practitioners in relation to the dominant anti-asylum seeking discourses. This chapter seeks to contribute towards filling this gap by analysing counter narratives provided by social work practitioners who were interviewed for this study. Andrews (2004, p. 1) defines counter narratives as 'the stories which people tell and live which offer resistance, either implicitly or explicitly, to dominant cultural narratives'. Implicit within this definition is the notion of a discursive struggle over meanings. This view of counter narratives is shared across various strands of discourse analysis. Laclau and Mouffe (1985) state that even though some narratives become the dominant or hegemonic discourses, such dominance is never complete. As a result, there is an ever-present ongoing struggle for hegemony among competing discourses. There is also a possibility that marginal oppositional discourses can ascend discursively to become dominant discourses.

The data analysis revealed a number of interpretative repertoires, which were utilised by social workers to provide counter narratives to the dominant ways of portraying asylum seekers. Within these interpretative repertoires various linguistic strategies were employed by social workers interviewed in their attempts to deconstruct and reconstitute asylum seeking service users.

Asylum seekers as misrepresented

The media and politicians were depicted as misrepresenting asylum seekers as a homogenous group. The media and politicians were blamed for their role in conveying a particularly negative view of asylum seekers. Respondents constituted asylum seekers as group that was used by politicians as a scapegoat for the country's social and economic malaise. Social worker 1's view was that in terms of politicians' representations of asylum seekers,

> a lot of politics is played out and [it] is about trying to please the masses. I think if they say, we have a number of immigration issues of immigrants coming across, again the press are very negative and put out very negative

DOI: 10.1057/9781137415042.0008

> images of people trying to come across here. I am sure it's a very small portion of displaced people who are here or seeking asylum or a small portion actually gets to claim asylum. I know a lot of people who are returned back to their countries of origin. (Social Worker 1)

The misrepresentation of asylum seekers by the politicians and the media was accounted for in terms of political expediency, that is 'to please the masses'. Politicians were also accused of playing the 'numbers game' in their representations of asylum seekers with a view to mislead the public. Research has demonstrated how politicians could manipulate statistics in order to put forward a case for a more restrictive asylum regime (van Dijk, 2000, Buchanan et al., 2003, Goodman, 2007). Social Worker 1 deployed the 'numbers game' as a discursive resource to great effect in his bid to constitute asylum seekers as deliberately misrepresented by politicians and the media. Quite a lot of the media coverage of the issue of asylum focuses on the question of the numbers within the United Kingdom and specifically how many continue to arrive in the country annually. Within the anti-asylum seeking discourses these numbers are then related, more often than not, to the cost to the British taxpayer. Other issues related to immigration and asylum such as the state of public services and crime are also contextualised by reference to the 'shock figures' (Buchanan et al., 2003).

By drawing on the 'numbers game' as a discursive resource, Social Worker 1 highlighted how statistics can be manipulated and presented as statements of fact. In doing so, he also provided a counter narrative to the anti-asylum seeking discourses that purport the United Kingdom is at risk of being overwhelmed by the numbers of asylum seekers. According to Social Worker 1, 'it's a very small portion of displaced people who are here'. Social Worker 13 also made a similar argument when she stated that a totally different picture emerges 'when you actually do the number crunching'. In 2010, the UNCHR indicated that only a relatively small proportion of the global refugee population of 15.5 million actually come to Europe annually (UNHCR, 2010a). The UNCHR estimates that 83 per cent of refugees stay in their regions of origin. In most cases refugees only manage to flee to neighbouring countries. This has resulted in the developing countries holding eight out of ten refugees. In 2010, Britain received the third highest number of asylum applications in Europe after France and Germany. In 2013, the number of asylum claims per capita in the United Kingdom was significantly lower than the average across

DOI: 10.1057/9781137415042.0008

Western Europe. The United Kingdom received 0.17 asylum applications per 1,000 compared to 0.91 across Western Europe (Blinder, 2014).

Asylum seekers as individual human beings

This strategy involves the provision of an alternative construct in which asylum seekers were presented first and foremost as individual human beings. The significance of this strategy is that it reclaimed the humanity that asylum seekers were stripped of in the dominant anti-asylum discourses:

> What I see on the news and local papers, I take with a pinch of salt because I know that, as I have just said, its sensationalism to sell papers. It incites anger, it incites hatred and I think it's very negative. When you are actually working with people, whether black or white, you are working with a person. It becomes much more. Umm; obviously you are working within boundaries in social work. You are working within a framework, you are also working with people, especially in this unit... I think we need to treat people as individuals, with respect and dignity, and I think that's what we hopefully do. (Social Worker 1)

It is also important to pay attention to some of the linguistic strategies that Social Worker 1 deployed in his attempts to re-constitute asylum seekers as individual human beings in this extract. Significantly, his formulation was argumentatively oriented as it was structured to undermine the constructs portrayed in the dominant anti-asylum seeking discourses. The social worker's formulation of asylum seekers began first by challenging what he believed to be erroneous views widely circulated by the press before he advanced his view that asylum seekers were indeed individuals, each with an individual story of their own. Thus, media depictions were shown as untrue and removed from the truth; they were dismissed as sensationalism, which could potentially worsen community race relations. Crucially, in order to give his formulation of asylum seekers credibility, Social Worker 1 strategically positioned himself as being in touch with asylum seekers on a regular basis by asserting that, 'When you are actually working with people, whether black or white, you are working with a person. It becomes much more'. This effectively positioned Social Worker 1 as having a more intimate knowledge and experience of working with asylum seekers. This particularly added to the persuasive force

DOI: 10.1057/9781137415042.0008

of Social Worker 1's account as emphasis was placed on actual first-hand and lived experiences within an institutional context where service users were treated equally regardless of race or nationality.

Asylum seekers as 'just like us'

The constitution of asylum seekers as individuals also has to be understood within the context of a counter discourse that seeks to challenge the notion of asylum seekers as those who are *not part of us* and as such present a threat to the British 'way of life' (Capdevila and Callaghan, 2008). Respondents sought to challenge this notion in their construction of asylum seekers by deploying various linguistic devices. Social Worker 5 constructed asylum seekers in the following terms:

> one would be in contact with and speaking to the families and seeing small children makes it much more personal. It's not the type that is coming in and taking our jobs and things like that. Quite often I find that the parents are intellectually trained. They want to work but they can't; and they are even willing to do jobs that a lot of Scottish people would sit at home and not do. There is this whole restriction on them on working here and we prefer to give them money. I would say it makes it more personal when I see children, I see them with their mums and dads, and I see them as people persecuted people at risk and it's even more shocking when you get to meet somebody personally. (Social Worker 5)

In this extract, Social Work 5 drew on the nuclear family as a discursive resource in order to portray asylum seekers in a positive light. The family unit was invoked as part of an endeavour to normalise asylum seekers. Thus Social Worker 5 referred to asylum seekers as 'families' consisting of 'mums', 'dads' and 'small children'. These 'parents' were also portrayed as 'intellectually trained' and motivated to work and contribute positively to their communities even if that meant doing jobs they were not trained for, including those 'jobs that a lot of Scottish people would sit at home and not do' (Social Worker 5). In order to complete the process of normalising asylum seekers, blame was then attributed to the asylum system, which was portrayed as preventing these families from being active members of their respective communities. Therefore, asylum seekers were presented as sharing with the rest of the British people the same putative social values that are held in high esteem

DOI: 10.1057/9781137415042.0008

within mainstream British society such as being family-centred and industrious. Constructing asylum seekers in this manner reinforced the argument that asylum seekers were just *like us* and as such should be treated like *we* would treat *our own*. Therefore, a counter narrative was provided to the dominant anti-asylum seeking discourses which tended to portray asylum seekers as socially deviant and having a very limited capacity to fully integrate into their new communities. In fact, the following extract from Social Worker 13's account was specifically worked up and oriented as a response to the claim that asylum seekers are socially deviant:

> I deal with human beings involved. I mean nobody is perfect. I have been involved with this young person and came across an individual who was an ex-asylum seeker who aren't completely honest about their motives, but you know what; if I was living on benefits maybe I would be slightly as well. I am not advocating that people should be breaking the law, but if you are struggling some people may, whilst I don't agree with them I can understand why they have done what they have done. But most of the people I have dealt with are very honest people who are frustrated and fed up. (Social Worker 13)

As a result, asylum seekers were constituted as having a lot in common with British citizens, the only difference being the legal status ascribed to them by immigration law. It is also worth drawing attention to how this construction of asylum seekers was formulated by Social Worker 13. This alternative construct has to be seen as oriented to the predominant depictions of asylum seekers as dishonest. What could be perceived at face value as socially deviant behaviour was instead reconstructed and rationalised as a *normal* response to an otherwise hostile environment. This was rationalised by the claim that faced with similar circumstances, Social Worker 13 would respond in the same way as asylum seekers; and such a response would not be perceived as socially deviant. It is also important to take note of how the press was dismissed as not depicting true events and disassociated from the actual human being. For this counter argument to be persuasive it was formulated and framed around the significance of lived experiences as Social Worker 13 argued that she dealt with the actual human beings involved in all this and the majority of whom were very honest people and if anything were struggling, frustrated and fed up and as such were far from the depictions in the press and politicians' speeches.

DOI: 10.1057/9781137415042.0008

Asylum seekers as deserving 'our' sympathy and care

As part of the attempt to provide an alternative construct, the respondents presented asylum seekers as individuals who deserved 'our' sympathy. Asylum seekers were depicted as vulnerable individuals:

> If you listen to the people's stories, to their life experiences, a lot of it is pretty traumatic, pretty horrendous to think how it would be to be in that position. The fact that they are given little money and little support and that is better than being back in their country of origin says loads; doesn't it? I guess working within [this team] has opened my eyes to a lot of these issues... Because you are working with people going through a difficult period, when people are destitute, have nowhere to live, when their lives are really low.... (Social Worker 1)

With reference to Australia, Every (2008, p. 657) has noted that, 'Whilst the equation of refugees with persecution is a common sense understanding, politically this link is being actively undermined by the introduced laws.' The same also applies to the United Kingdom and elsewhere in Western Europe where a culture of disbelief and suspicion of the motives of asylum seekers is firmly established. 'Persecution' as a reason for seeking asylum is fast losing its persuasive appeal. The idea that people who seek asylum in the United Kingdom are fleeing persecution has come under heavy attack mainly from anti-asylum discourses (Masocha and Simpson, 2011b). The term has increasingly been appropriated by the anti-asylum seeking lobby and is frequently referred to in calls to restrict even further the numbers of asylum seekers that can be granted protection. Respondents appeared to have an awareness of this and the futility of attempting to engage in this debate directly. Instead, they did so rather effectively without directly referring to the actual persecution that asylum seekers could have possibly suffered in their respective countries of origin. For instance, in the above extract, Social Worker 1 focused on the difficulties experienced by asylum seekers *within* the United Kingdom as a way of demonstrating that there are genuine reasons, why asylum seekers come into the United Kingdom to seek sanctuary. He posed the question, 'The fact they are given little money and little support and that is better than being in their country of origin says loads; doesn't it?' (Social Worker 1). By asking this rhetorical question, Social Worker 1 effectively undermined the argument that is advanced in anti-asylum discourses that

DOI: 10.1057/9781137415042.0008

asylum seekers come to the United Kingdom for economic reasons not because they are persecuted.

Therefore, asylum seekers were shown to be vulnerable not only because of what they experienced in their countries of origin and during the journey to the host country, but also due to the various forms of treatment that they then receive in the United Kingdom due to their precarious immigration status. Thus a discourse of 'lack of choice' emerges in which asylum seekers were constituted as having very limited choices in their lives as evidenced by the limited access to mainstream welfare provisions and limited opportunities to access some of the support systems and networks available to British citizens. This was attributed to factors such as persecution in their countries of origin and host countries as well as the harsh and restrictive legislation in host countries. The effect of this was that it evoked social work's moral obligations and duty of care towards asylum seekers.

The moral obligation and duty of care towards asylum seekers was further reinforced through constituting asylum seekers as struggling to meet their day-to-day needs and as such were deserving of a service:

> Asylum seekers have no access to public funds; have no access to funds. Big, big issues; where do they get food; where do you get your accommodation; when do you get it, how do you pay for it? So, you may be able to apply for accommodation within [the city], not through the public, but private sector, you are faced with; I might get the house, but how am I going to pay for it? If I can't work, how am I going to find the actual payment for rent, paying council tax, paying utility bills, and paying food bills? It's just a nightmare. Such a lot of pressure, a lot of issues, as I said earlier, which a lot of locals don't have because they are on benefits and get access to funds, access to housing; additional resources which asylum seekers don't have. (Social Worker 1)

Here, the use of rhetorical questions to emphasise the gravity of the situation and the appropriation of an asylum seeker's voice were relevant and significant linguistic strategies. The respondent invited the reader into an asylum seeker's introspection. The effect was to bring the reader into the 'nightmare' world of the asylum seeker and his innermost feelings. This way the observations made by Social Worker 1 were given a human voice; the asylum seeker who is struggling on a daily basis to make ends meet was allowed to speak for him/herself. The fact that this was used in conjunction with the rhetorical questions was very relevant as it served as a further illustration of the difficulties that

DOI: 10.1057/9781137415042.0008

asylum seekers face, the lack of choice in particular. This lack of choice was further evidenced through the use of contrasts. The difficulties that asylum seekers experience were contrasted with those of British citizens. The effect was to produce a counter narrative which showed asylum seekers as worse off than their counterparts as they lacked the safety net provided by mainstream benefits which served to further emphasise asylum seekers' vulnerabilities particularly to poverty and destitution.

Asylum seekers as a resource

Asylum seekers were constituted as a potential resource that could be utilised for the benefit of the nation. According to Social Worker 1:

> I think these are people with skills. They can work and want to work. We should support people to get into employment and stay there for a long time, or support them into college or university so that they can do some training whilst they are here and make good use of that time. At the moment, that is something that is not possible. I think we have got a lot of very talented people that we can actually use and support to go into the workplace and use their skills and talent, and yet current legislation and policies don't just provide that. A lot of these people do want to work. Umm; I think we should engage with that whilst they are waiting for a decision to be made.

Therefore, asylum seekers here were constructed as possessing the much emulated characteristics of the *good immigrant*. Capdevila and Callaghan (2008, p. 10) have noted that in anti-asylum discourses, *good immigrants* are often 'represented as hard working people *like us*, with whom we share a common vision of Britain and British-ness determined by a shared moral sense, shared cultural identity – not by birth ... but by a sense of common identification with the British way of life'. Within the dominant anti-asylum seeking discourses, asylum seekers are constituted as social deviants who do not fulfil this criteria of what constitutes a good immigrant. However, in their constructions of asylum seekers, respondents provided a counter narrative that demonstrated that asylum seekers in fact possessed most of these characteristics but this was either deliberately underplayed in anti-asylum seeking discourses or the existing legislation simply made it difficult if not impossible for asylum seekers to demonstrate that they possessed them.

DOI: 10.1057/9781137415042.0008

In order to accomplish this task of constituting asylum seekers as *good immigrants*, some blame work had to be undertaken in order to explain why these characteristics were not readily identifiable. Social Worker 1 brought into focus the discord between policy and the situation on the ground. One such discord was the fact that Scotland needs manpower due to its demographic composition yet it had asylum seekers who are willing to work but were being forced to be dependent on state support primarily due to political reasons. Through referring to specific cases that they had been involved with, respondents were able to illustrate this contradiction even more clearly. For instance, Social Worker 7 referred to cases in which she depicted asylum seekers as willing to work but forced into a life of dependency on state benefits:

> I worked with a family from Zimbabwe who were failed asylum seekers but had written to the Home Office for a legacy and they lived in England and came to [this city] and even though asylum seekers need permission to work he managed to work for the council in the local community library. He worked through there for a couple of years and it was only after he went for promotion that the council realised that they were employing someone who was an asylum seeker and had no legal kind of right to be in the country. He was working and contributing towards the community in the work he was doing but his immigration status says you are not allowed to work. And it came to the point where it became a case where the social work department are paying for their maintenance yet they were willing to work.

The effect of such a formulation was that it showed asylum seekers in a much more positive light. For instance, the motivation and desire to work and contribute positively to the community was given prominence by the fact that the asylum seeker in this case had to cheat the system to be a *good immigrant.* The picture that then emerged was one in which existing legislation and policies were portrayed as significant stumbling blocks. Through such a depiction, the responsibility for asylum seekers ending up being a burden to the taxpayer was shifted away from asylum seekers as it was explained in terms of what was portrayed as unreasonable constraints imposed by existing legislation and policies. Social worker 7 used this specific case to evidence her assertion that, 'Most of the asylum seeking families we have are willing to work and want to work and do any type of work and yet are not allowed and we are paying them money from a budget we don't have. I think it's ridiculous!' (Social Worker 7). This way case-talk was used as a discursive strategy to challenge the

DOI: 10.1057/9781137415042.0008

notion that asylum seekers were a drain on public resources. Research has demonstrated how this notion of asylum seekers as a drain on the public purse features predominantly in anti-asylum seeking discourses and how it is used to justify calls for tougher anti-asylum seeking legislation (Every and Augoustinos, 2008b, Gabrielatos and Baker, 2008, Masocha and Simpson, 2011b). Here the same notion was used to serve a different purpose. It was used to illustrate the fact that asylum seekers were viewed as a drain on public resources was primarily because of madness of the existing legislation. Thus the blame was shifted away from asylum seekers and put squarely onto the government.

Asylum seekers were also constructed as a source of cultural richness and diversity, which could contribute to racial tolerance:

> We need to have a mixture. I mean, having lived in London I noticed when I came to [this city] how white the population appeared having been used to such a mixed population. I think it's changing and I think it's good from kind of racial attitudes if you like. There is a lot of ways people can give; it's not about money. (Social Worker 5)

It is important to note that the overall effect of constructing asylum seekers in this way was that it also provided a counter argument to claims by politicians such as Anne Widdecombe and Michael Howard that immigration and asylum pose a significant threat to community relations, racial harmony, even national security and the British way of life. Thus a counter narrative was provided to the spectre of the *enemy within* (Fekete, 2004) as asylum seekers were constituted as a pool of human resources that could be tapped into to tackle some of the country's socio-economic problems.

Discussion

The subject position assumed in the counter narratives that were provided in this study were anchored in social work values of unconditional positive regard, acceptance as well as a quest for social justice. Interviewed practitioners portrayed social work as 'naturally' predisposed to be sympathetic towards asylum seekers. The respondents pointed out that by its very nature social work's role in society is to work with vulnerable members of society; and asylum seekers are one such group. Social work was constituted as having a duty of care towards asylum seekers

DOI: 10.1057/9781137415042.0008

due to the nature of asylum seekers' needs. Within these counter narratives, respondents unequivocally identified a moral duty towards this service user group. Respondents were also clear that this is a legitimate area of social work practice, which is compatible with the profession's commitment towards the social justice agenda. In providing alternative constructs of asylum seekers, counter narratives were able to portray asylum seekers as individuals with multiple identities. Respondents argued that the label of being an asylum seeker was only a part of their multiple identities. Such a portrayal challenged the predominantly negative monolithic constructs that are found in anti-asylum seeking discourses.

By engaging in these discursive acts, social workers were able to portray positively not only asylum seekers but also the profession as a whole. The social workers' dissenting voices were able to reject the inevitability and sense of 'expectation' that the kind of treatment asylum seekers receive was necessarily part of being an asylum seeker. By engaging in such discursive acts and taking up specific subject positions, the practitioners were able to foreground the humanity and individuality of asylum seekers.

What emerges clearly in this chapter is that for social work to make a significant difference to marginalised service users, the profession not only needs to have an in-depth understanding of its service user groups, but also needs to be actively involved in discursive acts that constitute those particular groups' identities. A key argument in this book is that the identity of asylum seekers is discursively accomplished. As such, there is a need to move beyond the simplistic and often essentialist typology of this service user group. Part of that imperative involved developing an understanding of how asylum seekers as a service user group are not a given but are constructed at various levels within society, and how the profession can assume a specific subject position within the ensuing discourses. According to Bishop and Jaworski (2003, p. 246), 'it is through discourse that social realities are articulated and shaped: people's perceptions of the world, their knowledge, and understanding of social situations, their interpersonal roles, their identities, as well as relationships between interacting groups of people ...'. Such an understanding can only be achieved through social work having a better understanding and actively participating in debates that potentially affect those whom they seek to attain social justice for. Significantly, social workers' counter

DOI: 10.1057/9781137415042.0008

narratives were analysed for their ability to *reconstitute* ideas on the topic of asylum seekers.

Admittedly, these positive views of asylum seekers presented in these narratives are not entirely unproblematic. For instance, these narratives can be criticised for failing to foreground asylum seekers' resilience and agency by constructing them as a vulnerable group as this masks the diversity of experiences among asylum seekers. It is also possible to criticise these narratives as somewhat assimilationist. Furthermore, it could be argued that asylum seekers are being constructed through neoliberal lens, which has the effect of treating them as commodities given the direct links that are made with the needs of the Scottish labour market. Nonetheless, these views represent a significant departure from the predominantly negative constructs of asylum seekers. Significantly, these social workers' narratives present a much more positive and optimistic view of practice with asylum seekers, which contrasts sharply with the deficit-laden perspectives presented in earlier studies of practice with asylum seekers (Collett, 2004, Humphries, 2004b, Mynott, 2005). The social workers' narratives also confirm the existence of supportive and innovative practice within a constrained environment (Fell and Fell, 2013, Kohli, 2006a).

DOI: 10.1057/9781137415042.0008

5
Construction of the Other

Abstract: *It can be argued that the professional and institutional contexts in which social workers are located make it very difficult for practitioners to blatantly express what might be perceived as oppressive, discriminatory, exclusionary, racist or prejudiced views. As such, there are very limited spaces for the expression of what might be construed as prejudiced or pejorative views in relation to a marginalised social group like asylum seekers. This chapter focuses on those narratives in which a minority respondent position was taken up. This chapter considers how negative representations of asylum seekers can be constructed within such limited spaces in the highly constrained and regulated ideological area of social work practice. The chapter demonstrates how such oppositional discourses are rhetorically organised and positioned within such restricted and heavily regulated contexts. It highlights discernible patterns in those* few instances *in which social workers' discourses negatively positioned asylum seekers and rationalised their continued marginalisation and/or exclusion from mainstream British society. This chapter takes an interest in the discursive strategies that were deployed to formulate such negative views of out-groups. It demonstrates how such negative formulations were presented as reasonable and justified while simultaneously protecting the interlocutor from potential accusations of being prejudiced.*

Masocha, Shepard. *Asylum Seekers, Social Work and Racism*. Basingstoke: Palgrave Macmillan, 2015.
DOI: 10.1057/9781137415042.0009.

 DOI: 10.1057/9781137415042.0009

Introduction

While the majority of the practitioners who were interviewed constituted asylum seekers in positive terms and were *predominantly* sympathetic in their formulations as the last chapter demonstrated, there were also a small number of narratives in which practitioners negatively constructed asylum seekers. This chapter focuses on those narratives in which a minority respondent position was taken up. Within any dominant discourse there are also spaces that can be taken up by oppositional discourses. Foucault (1998, p. 100) states that, 'We must conceive discourse as a series of discontinuous segments whose tactical function is neither uniform nor stable.' As such, there are always spaces for resistance, which can be taken up by oppositional pedagogies. This chapter considers how negative representations of asylum seekers can be constructed within such limited spaces in the highly constrained and regulated ideological area of social work practice. Within the social work profession there is a dominant narrative, which unambiguously aligns the profession with the vulnerable and marginalised social groups in society, and indeed most of the social workers that took part in the interviews for this study subscribed to that discourse. This dominant narrative is underpinned by social work values and ethics. It affirms explicitly the profession's commitment to the social justice agenda as well as anti-racist and anti-discriminatory practice (Dominelli, 2012, Parrott, 2009, Thompson, 2006). Thus, it can be argued that the professional and institutional contexts in which social workers are located make it very difficult for practitioners to blatantly express what might be perceived as oppressive, discriminatory, exclusionary, racist or prejudiced views. As such, there are very limited spaces for the expression of what might be construed as prejudiced or pejorative views in relation to a marginalised social group like asylum seekers.

It is therefore apposite that an analysis is undertaken to enable an in-depth understanding of how such oppositional discourses are rhetorically organised and positioned within such restricted and heavily regulated contexts. This chapter analyses discernible patterns in those *few instances* in which social workers' discourses negatively positioned asylum seekers and rationalised their continued marginalisation and/or exclusion from mainstream British society. Thus, this chapter takes an interest in the discursive strategies that were deployed to formulate such negative views of out-groups. It demonstrates how such negative

DOI: 10.1057/9781137415042.0009

formulations were presented as reasonable and justified while simultaneously protecting the interlocutor from potential accusations of being prejudiced.

The data analysis revealed a number of interpretative repertoires, which were drawn upon by a minority of the social workers to produce negative formulations of asylum seekers. Within these interpretative repertoires, a number of linguistic strategies were deployed by social workers to constitute asylum seeking service users in negative ways. As noted, the current socio-political climate, professional codes of conduct and ethics play a key role in stifling the expression of prejudiced sentiments within a social work practice context. This chapter illuminates how the cited social workers positioned their narratives and made use of a variety of rhetorical strategies. These strategies had the effect of inoculating the respondents from potential accusations of being prejudiced that could have arisen as a result of the negative formulations of asylum seekers. In so doing, it became possible to say the otherwise 'unsayable' (Every and Augoustinos, 2007b). The negative representations of asylum seekers were carefully formulated within the bounds of 'reasonable prejudice' (Billig 1988) and were oriented to ward off anticipated negative attributions in advance.

The cultural Other

A minority of social workers identified asylum seekers as the outsider on the basis of their culture. Research in the Netherlands has demonstrated how on the basis of culture immigrants were perceived by social workers as outsiders who did not fit into the available classifications (Scheppers et al., 2006). In this context, culture became a way of differentiating people or what van Dijk (2004) calls the 'cultural others'. Within the interviews conducted a few of the social workers deployed the term culture as a signifier of difference – a particular marker for ethnic minorities in the United Kingdom. Within their accounts, these respondents constituted asylum seekers as the cultural Other. Within that context, culture was portrayed in a largely negative sense. For instance, an asylum seeker's culture was portrayed as an impediment to social integration:

> I think, if I am going to be completely honest and not be PC again, I think there are aspects where at times people perhaps should realise that they are now in Britain and there are things perhaps that may not be acceptable in

DOI: 10.1057/9781137415042.0009

> their country, but are unfortunately the ways we do things here; and that can be hard if you come across it. Some of my colleagues here one day had to run out to the back because there was an Asian family beating up a fourteen year old boy and they said that was ok in their culture, but to us that is completely not acceptable in our country. (Social Worker 4)

The use of specific linguistic devices by Social Worker 4 here needs to be noted. Of particular importance is the strategic use of the ethnic marker 'Asian' as well as how this is deployed alongside the *us v them* binary in the account which effectively depicts the asylum seeking family in question as the cultural Other. Furthermore, the use of the phrases 'our country' and 'their country' has a relevant spatial illocutionary force, which effectively reinforces the fact that the asylum seeker was essentially an outsider. The spatial illocutionary effect was further reinforced by Social Worker 4's insistence that this was the way '*we* do things *here*'. Thus, in this context, culture was used as a relational and 'referential demarcator measuring the distance these Others stand in relation to the Caucasian mainstream' (Park, 2005, p. 21). By 'telling the case' (Pithouse and Atkinson, 1988), Social Worker 4 was able to persuasively mark the service users' behaviour as deviant and in sharp contrast with perceived normality. The moral character of the service users was made available through their actions as the case was narrated.

Thus within this particular narrative, the asylum seeking family was constructed as not like 'us' and not willing to acculturate to the hegemonic British ways of life. Such characterisation was not dissimilar to that depicted in anti-asylum discourses (Masocha and Simpson, 2011b). The asylum seeking family was depicted as demonstrating an unwillingness to acculturate to 'British' ways of life despite having made a conscious decision to live in Britain. The effect of this was that the depicted family became complicit in its social marginalisation.

Constitution of asylum seekers as economic migrants

Furthermore, asylum seekers were also constituted as economic migrants by a minority of social workers that participated in the research study. Although these few social workers concerned did not categorise asylum seekers as bogus, this was implicit in their depictions. The following extract from Social Worker 25 clearly conveyed the same notions, as in media and parliamentary discourses, which purport that the vast

DOI: 10.1057/9781137415042.0009

majority of asylum seekers are not genuinely seeking sanctuary from persecution but are rather motivated by economic motives:

> Of course that [asylum] is a route into a life of, you want to call it say privilege; but privilege is such a relative thing. But, it is privilege for somebody in Pakistan living in those conditions, you know. It's access to all those things; access to welfare benefits, education and other things you wouldn't have, and a lifestyle you wouldn't have, and a lifestyle you wouldn't be able to afford at home; the comfort of security and all the other things that the human condition craves. (Social Worker 25)

This depiction of asylum seekers made links between asylum seeking and economic migration and in the process cast a shadow on the reasons given by individuals for seeking asylum.

In order to persuasively construct asylum seekers as economic migrants, a number of linguistic devices were deployed. For instance, Social Worker 25 began his formulation of asylum seekers as economic migrants by stating that he did not necessarily consider himself as an expert in asylum seeking matters. However, he went on to state that he had vast experience working with this service user group and claimed that 'there won't be anybody within [the local authority] who has as much experience' and that he had 'witnessed the evolution of legislative changes in the last ten years' (Social Worker 25). Beginning his formulation with a disclaimer that he was not an expert but going on to stake a claim that he had unrivalled knowledge and experience is a rhetorical strategy – *diminution.* The effect of deploying this rhetorical device was that Social Worker 25 called attention to his expertise in this area of practice in the very act of disclaiming it.

Furthermore, prefacing his views with such a rhetorical device had the effect of positioning the social worker as a humble but expert respondent which made the negative formulation of asylum seekers which then followed somewhat credible, reasonable and more or less acceptable as it was presented as based on practice wisdom. The nature of practice wisdom as a knowledge base for social work has been a subject of extensive research (Chu and Tsui, 2008, Goldstein, 1990, Hudson, 1997, O'Sullivan, 2005). Practice wisdom can be defined as knowledge that has been gained through direct work with service users. As an episteme, practice wisdom is 'constituted intersubjectively and grounded in personal contexts and local sites' (Chu and Tsui, 2008, pp. 48–49) and relates to specific practice contexts. Thus, Social Worker 25's view that a

DOI: 10.1057/9781137415042.0009

lot of the asylum seekers were economic migrants was closely linked to both practice wisdom and what the social worker perceived as common knowledge:

> I think because of my experiences in working with asylum seekers for so long... it will be stupid to say there aren't out of the numbers that I have worked with a lot of people that are chancing it. Of course any system is open to exploitation. And I think what has happened in the UK's asylum policy is that a lot of people who are economic migrants are using asylum as means of entering the UK. (Social Worker 25)

Here it was significant that Social Worker 25 invoked shared knowledge as a linguistic strategy. In this formulation, what was advanced as a general rule and common knowledge (i.e. any system is open to exploitation) was evidenced using the social worker's practice wisdom derived from working with asylum seeking service users. Similarly, Social Worker 2 also related how from practice experience she knew that many economic migrants were posing as asylum seekers and how it was common knowledge that many of those that were presenting at airports were 'being advised before they arrived into the United Kingdom that the way to get into the United Kingdom and have some freedom of movement was to make an asylum claim' (Social Worker 2). By combining practice wisdom and what was generally presented as common knowledge, both Social Worker 25 and Social Worker 2 were able to persuasively portray the majority of asylum seekers as not genuine but in fact 'chancing it'. The effect of utilising practice wisdom and combining it with a body of knowledge that was presented as common knowledge was that it distanced these social workers away from the negative formulations of asylum seekers that were then made. The negative formulations were then presented as not emanating from the personal prejudices that the individual social workers possibly had. Instead it was the evidence that was leading these social workers in that direction.

Social Worker 24 also drew on practice wisdom in his formulation of asylum seekers as economic migrants. According to the social worker, before working with asylum seekers, he held very naïve views. He prefaced his formulation of asylum seekers with an explanation of his political views. He unambiguously identified himself as a liberal and in support of people's rights as enshrined in the European Convention of Human Rights and declared his support for the right to seek asylum. Prefacing his formulation of asylum seekers with his political views

DOI: 10.1057/9781137415042.0009

was a linguistic move that was relevant and important. It pre-empted and offered a rebuttal to any potential accusations of prejudice that may be made as a result of the adverse formulation of asylum seekers that then follows. The negative formulation that was made of asylum seekers by Social Worker 24 was attributed to the nature of asylum seekers themselves; otherwise the interlocutor himself was predisposed to be sympathetic to asylum seekers given his political views and beliefs.

What we are doing for them they would not be able to do it for us

In addition, the long-standing British tradition of generosity towards the less fortunate was also drawn upon as a discursive resource in the negative formulation of asylum seekers. For instance, Social Worker 24 constructed asylum seekers in the following terms:

> In the end I would be interested in us going to another country and see how we will be treated by their social services departments. I don't know; different countries have different systems. In certain countries they don't have the NHS like we have here. There are aspects of our healthcare and there are cases of people coming here suddenly being diagnosed with serious illnesses and are unable to go back to their countries because they will not be treated for those illnesses. I don't think for a minute they didn't know before they came. Ok, that's unfair because it doesn't apply to everybody but yeah I mean they are clever. Well done. Absolutely, you can't help but give them credit for being clever about it. They didn't go to America because they would have to pay there. I mean there are good things about our country in that respect and we would never leave anybody being ill when we can help them. (Social Worker 24)

Social Worker 24 used British generosity as a broader context for understanding the nature of asylum seekers who come to the United Kingdom. In order to achieve this, categories and the moral character of asylum seekers were invoked and inferred upon as a way of emphasising the type of client that social services were looking after. Thus categories of asylum seekers that are found in wider anti-asylum seeking discourses were drawn upon. In Social Worker 24's account, categories of asylum seekers as a '*sneaky*', '*diseased*' social group that was *a burden* to the welfare system were invoked. With reference to Australia, Pickering (2001) argues that constructing asylum seekers through metaphors of disease has the effect of depicting

DOI: 10.1057/9781137415042.0009

them as posing a credible threat to the host society. Thus, the linking of asylum seekers with diagnoses of serious illnesses in Social Worker 24's narrative had the overall effect of depicting asylum seekers as the undesirable Others whose expulsion is necessary for the health of the nation.

The notion of Britain as a destination of choice (Crawley, 2010) for asylum seekers was also invoked within this narrative as a way of casting doubt that these asylum seekers were in the United Kingdom due to persecution in their countries of origin. It was significant that Social Worker 24 immediately followed up this categorisation of asylum seekers by emphasising British generosity by asserting 'I mean there're good things about our country in that respect and we would never leave anybody being ill when we can help them' (Social Worker 24). The importance of making that assertion was that in spite of the depicted moral character of asylum seekers, the British society still went out of its way to look after them. Therefore, by implication, asylum seekers who 'evidently' were *undeserving* (Sales, 2002b) of the services that they were receiving were not supposed to complain about service standards. In any case, the quality of care received by asylum seekers in the United Kingdom cannot be matched in their very own countries of origin. Therefore, taking the focus away from the shortfalls in current practice through highlighting the moral character of asylum seekers and the fact that their respective countries of origin would not be able to provide comparable levels of care if roles were reversed was an effective linguistic move in defending practice.

An important thread that runs throughout this account was also the use of the oppositional binary *us/we v them/they*. Fairclough (1989, p. 127) discusses the use of the inclusive '*we*' in a *Daily Mail* editorial ('*We*' cannot let *our* troops lose the edge ...), noting that the editorial was making a claim to speak for others including its readers and all British citizens and Social Worker 24 did the same here. Fairclough (1989, p. 128) argues that the effect of this linguistic strategy is that it 'serves corporate ideologies which stress the unity of a people at the expense of recognition of division of interest'. Thus, in the context of the above extract, the inclusive *we* was used to construct a consensus and unity in opposition against a constructed *them*, who were portrayed as taking advantage of *our* generous nature. Furthermore, important parts of this strategy also involved positive self-representation and negative Other representation. The respondent attributed positive and negative lexical items for *us* and *them*, respectively. Those designated as *us* were characterised as generous

DOI: 10.1057/9781137415042.0009

people who took care of the needy and would never leave anybody being ill when they could help them. Those in the out-group *them* were characterised as sneaky, diseased and seeking asylum under false pretences.

Having it better than our own

One respondent refuted the view that asylum seekers were unjustly treated in the United Kingdom. According to Social Worker 18:

> It can be quite difficult for social work giving asylum seeking families £120 per week, and then going out and seeing somebody whose kids are lying on the floor, don't have any beds and things like that. I am not saying that these people are not entitled to their money but when you are giving £120 per week to them and you can't give this family anything and they are living in dire poverty and mum is struggling to find food weekly, and for whatever reason we can't give her money. So we are doing it for them but we can't do it for ours. It's just unfortunate. (Social Worker 18)

Therefore, the argument that was advanced in this vivid description was that if anything asylum seeking service users were better off than their British citizen counterparts and that was primarily because they received preferential treatment at the expense of British citizens. The persuasive strength of this account came from its social and spatial location of asylum seekers as the outsiders – *them* – who do not belong *here*, are not part of *us* and, as such, are undeserving of the preferential treatment that they receive at the expense of *our own*. The persuasive strength of the account also emanated from its ability to invoke a powerful category of a witness to the perceived injustice: the poverty-stricken British family. This was achieved through the extreme case formulation of poverty that was presented through the depiction of British kids 'lying on the floor' with no beds, living in dire poverty and British mums 'struggling to find food' with social work not being able to do anything to help. This was particularly effective in conveying a strong sense of injustice. Such a formulation underscored the argument being made here that despite being the Outsider, asylum seekers were better off than British citizens.

Discussion

What emerged clearly from the analysis of the presented interpretative repertoires was that asylum seekers were portrayed within this dissenting

DOI: 10.1057/9781137415042.0009

discourse as a problematic out-group. This was accomplished through the process of rhetorical Othering in which the oppositional binary *us v them* is deployed to great effect. The process of rhetorical Othering was characterised by the positive portrayal of groups and individuals who were subsumed under the first person plural pronouns *us* and *we*; and the simultaneous marginalisation of those groups and people that were designated as *they* and *them*. According to Baker (2006, p. 16) 'such oppositions are typical of ideologies in that they create an inherent need to judge one side of the dichotomy as primary and the other as secondary, rather than thinking that neither can exist without the other'. Within that context, positive attributes of those designated as *us/we* were also juxtaposed or contrasted with those classified as *they/them* as illustrated. Coe et al. (2004) have argued that discourses that consistently deploy binaries privilege one over the other. In the case of the presented narratives, once the 'two' sides were constructed, they were treated unequally especially given the evaluative nature of the discourse. As part of that process of establishing these binaries 'the social construction of evil is necessary for the construction of good' (Achugar, 2004, p. 317). Within the narratives presented in this chapter, this took the shape of those in the *we/us* in-group (social workers/British society) being constructed in a positive light while *them* or the Others out-group (asylum seekers) were constructed in a negative light.

Similar to the anti-asylum seeking discourses discussed in Chapter 3, the use of the oppositional binary *us/them* was also important in framing the discourse of nations and nationhood, which was also a crucial component in the ways in which asylum seekers were constructed. A recurring theme across the presented narratives was how asylum seekers were framed as the 'outsider' that had the potential to destabilise, undermine or threaten the British communities and their ways of life. The seemingly benign deployment of notions of nations and nationhood as discursive resources in fact specifically caters for the interests of those in the in-group. The presence of these banal forms of nationalism (i.e. the extensive deployment of *us/we* binary) in the social workers' narratives presented in this article served to normalise as a given the importance of the nation, nation state and nationhood. These notions were of particular significance when explicitly worked up in discourses as they further emphasised the position of asylum seekers as the out-group, the Other.

Admittedly, the sample that is the focus of this chapter is a rather small one, which may potentially be perceived as adversely impacting on the

DOI: 10.1057/9781137415042.0009

extent to which the findings themselves can be generalised. As noted, the highly constrained nature of social work discourse makes it improbable that larger samples can be drawn upon. There is a need to emphasise that the insights that this small sample provides are significant as they offer opportunities for an in-depth nuanced understanding of oppositional pedagogies within a highly constrained ideological practice. In fact, this chapter illuminates a range of discursive practices that were deployed to realise specific rhetorical accomplishments, which may be generalisable across contexts especially in terms of their effects (Goodman, 2008). It is also crucial to highlight that within discursive psychology, the positivist concerns with generalisability are not accepted as of particular significance. The idea that research should closely capture '*the real world*' that can be replicated is problematic given the epistemological and ontological assumptions that underpin discursive psychology, which maintain that the world is discursively constructed (Hardy and Phillips, 2002). In fact, the chapter illuminates the explanatory potential of the analytical frameworks utilised, including their ability to provide new explanations (Potter and Wetherell, 1987), which may result in a better understanding of how exclusionary discourses may be constructed and positioned within social work discourses. As such, the minority respondent views that were expressed in the presented narratives deserve the attention they have been afforded in this chapter primarily due to the important insights they provide on the nature of exclusionary discourses within restricted ideological spaces.

The narratives presented in this chapter function precisely to negatively formulate and position asylum seekers as an out-group and in many ways rationalise this group's continued marginalisation and exclusion from mainstream British society. Of particular concern for social work is how such discourses can still manage to articulate exclusionary and discriminatory sentiments in spite of the prevailing social norms and taboos relating to the expression of prejudiced views, the predominance of anti-discriminatory and anti-racist perspectives in practice as well the existence of professional codes of conduct and ethics. It is therefore important that social work is adept in identifying and responding to the shifting nature and parameters of exclusionary discourses. As a first step this requires the development of a more nuanced understanding of how exclusionary tendencies can be articulated in contemporary practice as illustrated in this chapter. This can be achieved through embedding in the social work curriculum and practice models a critical understanding

DOI: 10.1057/9781137415042.0009

of the central role that language plays not only in the constitution of subjectivities but also in the legitimation of existing social relations and inequalities. The discursive approach utilised in this book can help social work to 'explicate the precise manner in which people articulate a complex set of positions that blend egalitarian views with discriminatory ones' (Every and Augoustinos, 2007b, p. 138). As illustrated, the principles of liberalism, equality, fairness and justice are not only inferred and drawn upon as discursive resources, but also actively pressed into the service of what is predominantly an exclusionary and discriminatory discourse. Thus, the extent to which social work research that utilises discourse analysis can contribute to an enhanced practice and in particular to the understanding of the complexities of contemporary prejudice and discrimination can hardly be overemphasised. The imperative to have such an in-depth understanding is quite clear given social work's orientation towards the social justice agenda.

Developing social work professionals' sensitivity to their own discursive practices will result in practitioners developing alertness to and an insight into the consequences of their use of language (Hall et al., 2006). This could go a long way towards the development of a culturally sensitive social work practice. Significantly this could also contribute towards the development of enhanced anti-racist frameworks that are not exclusively reliant on race and biological categorisation (Masocha and Simpson, 2011b) but focus on the ways in which discourses of exclusion are constituted and articulated through paying particular attention to language use. Such an addition to the current anti-racist and anti-oppressive frameworks would sufficiently respond to the shifting nature and parameters of exclusionary discourses.

DOI: 10.1057/9781137415042.0009

Conclusion

Abstract: *This chapter reviews the main themes of this book and discusses the implications for social work practice.*

Masocha, Shepard. *Asylum Seekers, Social Work and Racism*. Basingstoke: Palgrave Macmillan, 2015.
DOI: 10.1057/9781137415042.0010.

This book has illustrated that the politicians and media are major purveyors of information on asylum seekers that is available to social work professionals as they attempt to make sense of this service user group. These macro discourses deploy a range of linguistic devices in their attempts to construct and deconstruct asylum seekers. These constructs, particularly within the anti-asylum seeking discourses, are achieved by the deployment of vivid lexicalisation within the framework of *us/them* bifurcation in order to depict asylum seekers as the Other. A key theme that permeates through all the chapters of this book is the manner in which xenoracism is generated, inflamed and worked up in various contexts including social work. However, one of the implications of the claim that the media and politicians are responsible for driving racism is that the public are not held responsible for their opinions and behaviour. The effect of such a formulation is that it problematises the media and politicians while at the same time excusing the British public as a whole as not really xenoracist. Such an assumption would be based on the belief that Britain is not really a racist society. However, the alternative understanding that is suggested and advocated for in this book is that in relation to immigration and asylum, Britain is a society in which xenoracism is embedded in social institutions and social practices. Arguably such a perspective has more radical implications and presents a much wider scope and opportunities for effecting change especially from a social work and advocacy perspective. An understanding of how xenoracism permeates through some of the media and political representations of asylum seekers such as the ones offered in this book will provide an irritant to social work's enforcement role. Such a critical understanding of how asylum seekers are constructed in discourses that are historically positioned and act as additional frames of reference for social work practitioners is essential for enhancing practice and will certainly address the concerns about social workers' complicity in the marginalisation of asylum seeking service user groups. Such a perspective based on the deconstruction of the language that is used to portray asylum seekers not only results in increased awareness and in the further development of a critical approach to practice but also complements current policies and efforts aimed at realising a culturally sensitive practice.

As illustrated, the fact that newspapers can offer two opposing interpretations on the same story involving asylum seekers or the fact that politicians may offer two different takes on the same piece of legislation is not merely a crude matter of *bias* involving the conscious

DOI: 10.1057/9781137415042.0010

use of rhetorical devices to achieve a specific effect. Significantly, the difference of opinions, interpretation and presentation of the story or piece of legislation also signifies the embedding of an ideological point of view in social practice. As Fowler (1991, p. 12) argues, the concept of bias 'assumes the possibility of genuine neutrality, of some new medium being a clear undistorting window. And [that] can never be ...'. In fact, Fairclough (1995) argues that the systems of values and beliefs – ideologies – that are expressed in language are far more subtle and embedded in the ways in which language is used. For instance, Chapters 2 to 5 have highlighted that the 'choice' of language structure over another has a much more covert ideological influence than just being a simple case of bias or simply a case of using a specific rhetorical device like hyperbole or extreme case formulation. Throughout these chapters, the use of subjective personal pronouns *we/us* and *they/them* in an oppositional binary is an ideologically influenced construct that has been shown to have significant implications in the establishment of asylum seekers as an out-group and in most cases legitimates their subsequent treatment. According to Sonwalker (2005, pp. 263–264), 'These terms are apparently simple notions; they are invoked in simple conversation and they figure often in newspaper headlines. But they reify a deep rooted and complex structure of values, beliefs, themes, and prejudices prevailing in a socio-cultural environment.' The book has highlighted how the *us/them* bifurcation is deployed as a framework by the media and politicians in the formulation and justification of exclusionary tendencies and policies. Chapter 5 has also shown the same framework being drawn upon as a discursive resource by a minority of the respondents to articulate cultural differences and constitute asylum seekers as the cultural Other with similar effects. Therefore, this book demonstrates that it is only through a deconstruction of the ways in which language is used in the portrayal of asylum seekers that a much more in-depth understanding of the underlying ideologies can be realised. Yet the use of such an approach in social work research has remained very limited. It is therefore important that such a perspective is promoted within social work research and that practitioners are conversant in the manner in which language is used to construct subjectivities and mediate practice. This also means that social workers need to be aware of their own discursive practices and their effects. This imperative can hardly be overemphasised especially given the centrality of language and discourse to social work itself.

DOI: 10.1057/9781137415042.0010

Furthermore, categorisation is a powerful political and rhetorical strategy that has far-reaching consequences for the participants in the asylum seeking debate and practitioners working with asylum seekers as they attempt to impose their own systems of classification. The ways in which asylum seekers are classified has serious implications on the ways in which as service users they are subsequently treated. For instance, it is significant to draw attention to how category constructions in anti-asylum seeking discourses in Chapter 2 focus attention on asylum seekers' legitimacy and how in the process this draws attention away from how best asylum seekers can be helped as a particularly vulnerable social group within the British society. Conversely, category entitlements are used in Chapter 4 to direct attention to how asylum seekers are in fact treated and how best they can be helped. Again, this underscores the imperative for the profession to be actively involved in discourses relating to those who are targets of social work intervention.

The perspective that is adopted in this book, which pays particular attention to the specific ways in which language is deployed to enact social and cultural identities, has potential to result in real gains being achieved within social work practice with asylum seeking service users, especially from an advocacy standpoint. It has the potential to successfully challenge and change key concepts that are used to portray asylum seeking service users. Fairclough (1995, p. 55) is of the view that discourse can be used in 'creative ways' to transform social practice. For instance, concepts can be defined as the categories, relationships and theories through which we conceptualise the world and relate to one another. Given that the meaning of a concept is dependent on prevailing discourses and that an understanding of the world depends on these concepts, it therefore follows that engagement with the prevailing discourses is a necessary component of a politically transformative social work practice. This is because, as Fairclough (1995) notes, as political and social practices, discourses have the capacity to establish, sustain and change power relations and the collective identities of groups such as asylum seeking service users. Through participating in such discursive acts social work has the potential to change the world as it is understood. Through advocacy, social work can assume a leading role not only in terms of actively participating in the discursive acts but also in striving for social justice for this service user group. For instance, as the concept of asylum seeker is applied to the individual by the immigration service or social services, it significantly changes the way the individual is then

DOI: 10.1057/9781137415042.0010

perceived and specifically has a direct impact on the sets of welfare apparatus and material practices that are then invoked. For instance, within social services, while a resident or citizen gains access to welfare benefits and services subject to fulfilling eligibility criteria, an asylum seeker is automatically barred from accessing mainstream welfare benefits and services, and consequently becomes the responsibility of the Asylum Support Service. As such, changing a concept may have a fundamental impact on the specific ways in which an object of knowledge is then socially accomplished. It is the contention of this book that it is at this level that discourse can potentially have the greatest impact on the social world and could also be of particular interest to social work practitioners especially from an advocacy perspective.

Given the current restrictive legislative and policy frameworks social workers find themselves operating within, social workers can largely depend on their social work values in their quest for social justice for this otherwise disenfranchised group of service users. The importance of social work values when working with asylum seekers can hardly be overemphasised. It is particularly important that social workers continuously reflect on their personal prejudices and ensure that these do not impede on their effectiveness when working with this group of service users. Social workers are products of the same society, which is laden with all forms of prejudices particularly xenoracism; and social workers may read the same tabloids that constantly barrage society with negative constructs of asylum seekers and are therefore not immune to the existing discourses.

Within social services, there is a general lack of an in-depth understanding of the needs of asylum seeking service users and how best to engage them as evidenced by the respondents' lack of confidence when dealing with the service user group. Having an understanding of how and why asylum seekers in general are portrayed in various ways, most of which are predominantly xenoracist, will certainly make practitioners more confident in taking up such cases. More importantly, having such an understanding will enable social work as a profession to engage more effectively in debates around its roles and remit. A critical understanding of how asylum seekers are constructed in some media and political discourses and how this is underpinned by xenoracism will help social work to adopt a more active role in advocating for social justice for this service user group. Such an understanding can certainly be the basis for the transformation of social work as a profession.

DOI: 10.1057/9781137415042.0010

The ever-present tensions and dilemmas endemic within social work practice with asylum seekers are quite apparent from the practitioners' narratives that have been presented in this book. Social work derives its legitimacy as a profession from its values of providing for the needy and protecting the vulnerable, yet social workers are required to exclude the very people they are supposed to protect and care for on the basis of their immigration status. It is this imperative that represents a substantial ethical dilemma and obstacles to good and effective practice (Duvell and Jordan, 2002). Social Worker 4 clearly expressed concern and discomfort over the enforcement role she assumed in her work with an unaccompanied minor and stated that she felt that this went against 'every grain' (Social Worker 4) of what social work is. At worst, social workers can easily become gatekeepers to services separating the 'deserving' from the 'undeserving' (Sales, 2002a). At the same time, this does not imply that at some point in history the social work profession was invariably a caring profession. Banks (2001, p. 16) recognises the ambivalent role that social workers have always played in society both 'as expressing society's altruism (care) and enforcing societal norms (control)'.

Working with asylum seekers is one of those areas of social work practice from which it can clearly emerge as a caring profession that strives for social justice. This can be achieved by effectively challenging the state's attempts to subjugate the profession. Practitioners who participated in this study attributed the difficulties they experience in their work with asylum seeking service users to the fact that social work has lost its voice. In fact, the struggle between the state and the profession is historical and indeed it is a 'continuous struggle that is an important dynamic in the on-going construction of social work' as a profession that is caring and predominantly driven by humanitarian motives (Humphries, 2004b, p. 31). In the words of Powell (2001, p. 161), 'Social work's capacity to survive depends upon its legitimacy as an authentic "humanising voice", rather than simply a conservative profession conveniently wrapping itself in the rhetoric of the market.' Therefore, as far as working with asylum seekers is concerned the profession needs to assume a radical stance and meet the existing challenges head on, even if this means challenging the very institutions that pay the wages:

> The profession has a choice to make a new moral effort, to find its anger about the plight of the poor, to engage its knowledge about the sources of inequality with a new sense of imperative and urgency. Asylum seekers and others subject to immigration controls are a dramatic example of the many

> regarded as undeserving, excluded, non-citizens, worthy only of derision, abysmal treatment and ultimate expulsion... It is time for a reconstruction of social work that draws on its radical historical strand and interprets this within a globalized and marketised context. The first step is a commitment to understanding the contemporary role of social work in this wider framework, leading to a praxis that involves action for change and the beating of a new drum. (Humphries, 2004a, pp. 39–40)

Utilising research that is informed by a discursive approach in its analysis will certainly enable social work to achieve this goal and maintain its orientation towards the social justice agenda. The antithesis to this is unbearable to contemplate as it is equivalent to enforcement counselling devoid of a social justice agenda, and can only result in a sterile and depoliticised practice.

More importantly, developing social work professionals' sensitivity to their own discursive practices will result in practitioners developing alertness to and an insight into the consequences of their use of language (Hall et al., 2006). This could go a long way towards the development of a culturally sensitive social work practice. Significantly this could also contribute towards the development of enhanced anti-racist frameworks that are not reliant on race and biological categorisation but focus on the ways in which discourses of exclusion are constituted and articulated through paying particular attention to language use. Such an addition to the current anti-racist frameworks would sufficiently respond to the shifting parameters of exclusionary discourses demonstrated in this book.

DOI: 10.1057/9781137415042.0010

References

1962. Commonwealth Immigrants Act 1962. London: HM Stationery Office.
1968. Commonwealth Immigrants Act.London: HM Stationery Office.
1971. Immigration Act. London: HM Stationery Office.
2014. Immigration Act. Norwich: HM Stationery Office.
1999. Immigration and Asylum Act 1999. London: The Stationery Office Ltd.
2002. Politicians are not doing enough to protect us. *The Express*, 13 August 2002.
2004. HIV tests for immigrants vital to safeguard the NHS. *The Express*, 29 July 2004.
2008. Caring for the wrong people. *South Wales Echo*, 3 September 2008.
Achugar, M. 2004. The events and actors of 11 September 2011 as seen from Uruguay: Analysis of daily newspaper editorials. *Discourse & Society*, 15, 291–320.
Andreou, A. 2013. Tweeting arrests of 'illegal immigrants' is a new low for the Home Office, *The Guardian*, Friday 2 August 2013, http://www.theguardian.com/commentisfree/2013/aug/02/immigration-offenders-home-office-tweeting [accessed 21 August 2013].
Andrews, M. 2004. Memories of mother: Counter narratives of early maternal influence. *In:* Bamberg, M. & Andrews, M. (eds) *Considering counter narratives: Narratives, resisting and making sense*. Philadelphia: John Brenjamins Publishing Company.
Augoustinos, M. 2001. History as a rhetorical resource: Using historical narratives to argue and explain.

In: Mchoul, A. & Rapley, M. (eds) *How to analyse talk in institutional settings*. London: Continuum.

Augoustinos, M. & Every, D. 2007. The language of 'race' and prejudice: A discourse of denial, reason, and liberal-practical politics. *Journal of Language and Social Psychology*, 26, 123–141.

Augoustinos, M., Tuffin, K. & Every, D. 2005. New racism, meritocracy and individualism: Constraining affirmative action in education. *Discourse Society*, 16, 315–340.

Bailey, O. G. & Harindranath, R. 2005. Racialised othering. *In:* Allen, S. (ed.) *Journalism: Critical issues*. Berkshire: Open University Press.

Baker, P. 2006. *Using corpora in discourse analysis*. London: Continuum.

Bale, T. & Hampshire, J. 2012. Immigration policy. *In:* Heppell, T. & Seawright, D. (eds) *Cameron and the Conservatives: The transition to Coalition Government*. Basingstoke: Palgrave Macmillan.

Bale, T., Hampshire, J. & Partos, R. 2011. Having one's cake and eating it too: Cameroon's Conservatives and immigration. *The Political Quarterly*, 82, 398–406.

Banks, S. 2001. *Ethics and values of social work*. London: Palgrave Macmillan.

Barclay, A. & Ferguson, I. 2002. *Seeking peace of mind: The mental health needs of asylum seekers in Glasgow*. Sterling: University of Sterling, Department of Applied Social Science.

Barker, M. 1981a. *The new racism: Conservatives and the ideology of the tribe*. London: Junction Books.

Barker, M. 1981b. *The new racism. Conservatism and the ideology of the tribe*. Londres: Junction Books.

Batty, D. 2013. Home Office tactics in illegal crackdown prompt Twitter storm, *The Guardian*, 2 August 2013, http://www.theguardian.com/technology/2013/aug/02/twitter-storm-home-office-illegal-immigration [accessed 21August 2013].

Billig, M. 1988. *Ideological dilemmas: A social psychology of everyday thinking*. London: Sage.

Billig, M. 1995. *Banal nationalism*. London: Sage.

Bishop, H. & Jaworski, A. 2003. 'We beat 'em': Nationalism and the hegemony of homogeneity in the British press reportage of Germany versus England during Euro 2000. *Discourse & Society*, 14, 243–271.

Bleasdale, L. 2008. Under attack: The metaphoric threat of asylum seekers in public-political discourses. *Web Journal of Current Legal Issues*, 1, 1–17.

DOI: 10.1057/9781137415042.0011

Blinder, S. 2014. *Migration to the UK: Asylum*, Migration Observatory briefing, COMPAS, University of Oxford, UK, July 2014 avalaible on: http://migrationobservatory.ox.ac.uk/briefings/migration-uk-asylum [accessed 24 October 2014].

Bloch, A. 2000. A new era or more of the same? Asylum policy in the UK. *Journal of Refugee Studies*, 13, 29–42.

Bloch, A. & Schuster, L. 2002. Asylum and welfare: Contemporary debates. *Critical Social Policy*, 22, 393–414.

Blommaert, J. 2005. *Discourse: A critical introduction*. New York: Cambridge University Press.

Briskman, L. & Cemlyn, S. 2005. Reclaiming humanity for asylum-seekers. *International Social Work*, 48, 714–724.

Buchanan, S. 2001. *What's the story? Sangatte: A case of media coverage of asylum and refugee issues*. London: Article 19.

Buchanan, S., Grillo, R. & Threadgold, T. 2003. *Whats the story? Results from research into media coverage of refugees and asylum seekers in the UK*. London: Article 19.

Capdevila, R. & Callaghan, J. E. M. 2008. 'It's not racist. It's common sense'. A critical analysis of political discourse around asylum and immigration in the UK. *Journal of Community & Applied Social Psychology*, 18, 1–16.

Card, D., Dustmann, C. & Preston, I. 2005. *Understanding attitudes to immigration: The migration and minority module of the first European Social Survey*. London: Centre for research and analysis of migration.

Carter, B., Harris, C. & Joshi, S. 1987. *The 1951–55 Conservative Government and the racialisation of black immigration*. Coventry: Centre for Research in Eethnic Relations, University of Warwick.

Cemlyn, S. & Briskman, L. 2003. Asylum, children's rights and social work. *Child & Family Social Work*, 8, 163–178.

Chantler, K. 2011. Gender, asylum seekers and mental distress: Challenges for mental health social work. *British Journal of Social Work*, doi: 10.1093/bjsw/bcr062. First published online: 21 June 2011.

Charteris-Black, J. 2005. *Politicians and rhetoric: The persuasive power of metaphor*. Basingstoke: Palgrave Macmillan.

Chu, W. C. K. & Tsui, M. S. 2008. The nature of practice wisdom in social work revisited. *International Social Work*, 51, 47–54.

Coe, K., Domke, D., Graham, E. S., Lockett, J. S. & Pickard, V. W. 2004. No shades of grey: The binary dscourse of George W. Bush and an echoing press. *Journal of Communication*, 54, 234–252.

Cohen, P. 1999. New ethinicities, old racisms. *In:* COHEN, P. (ed.) *New ethnicities, old racisms*. London: Zed Books.
Cohen, S. 1994. *Frontiers of identity: The British and the others*. London: Longman.
Cohen, S. 1996. Anti-semitism, immigration controls and the welfare state. *In:* TAYLOR, D. (ed.) *Critical social policy: A reader*. London: Sage.
Cohen, S. 2002. The local state of immigration controls. *Critical Social Policy*, 22, 518–543.
Cohen, S. 2003. *Noone is illegal*. Stoke-on-Trent: Trentham Books.
Collett, J. 2004. Immigration is social work issue. *In:* HAYES, D. (ed.) *Social work, immigration and asylum: Debates, dilemmas and ethical issues for social work and social care practice*. Philadelphia: Jessica Kingsley.
Conservative Party. 2001. Conservative Party General Election Manifesto: Time for common sense, London: Conservative Party.
Conservative Party. 2010. *Invitation to joing the government of Britain. The Conservative Manifesto 2010*. London: Conservative Party.
Coole, C. 2002. A warm welcome? Scottish and UK media reporting of an asylum-seeker murder. *Media, Culture & Society*, 24, 839–852.
Crawley, H. 2010. *Chance or choice? Understanding why asylum seekers come to the UK*. London: Refugee Council.
Daily Mail. 1998. The good life on asylum alley, 6 October 1998.
Darylmple, J. & Burke, B. 2006. *Anti-oppressive practice: Social care and the law*. Maidenhead: Open University Press.
Dixon, J. & Wade, J. 2007. Leaving care? Transition planning and support for unaccompanied young people. *In:* Kohli, R. & Mitchell, F. (eds) *Working with unaccompanied asylum seeking children: Issues for policy and practice*. Basingstoke: Palgrave Macmillan.
Dominelli, L. 1988. *Anti-racist social work: A challenge for white practitioners and educators*. London: Macmillan.
Dominelli, L. 2008. *Anti-racist social work*. Basingstoke: Palgrave Macmillan.
Dominelli, L. 2012. Anti oppressive practice. *In:* GRAY, M., MIDGELY, J. & WEBB, S. (eds) *The Sage handbook for social work*. London: Sage.
Dover Express. 1998. We want to wash the dross down the drain, 1 October 1998.
Dovidio, J. F. & GAERTNER, S. L. 1986. *Prejudice, discrimination, and racism*. Orlando: Academic Press.
Dumper, H., Malfait, R. & Scott-Flynn, N. 2006. *Mental heath, destitution and asylum seekers – A study of destitute asylum seekers in the dispersal*

DOI: 10.1057/9781137415042.0011

areas of the South East of England. London: National Instutite of Mental Health in England.

Duvell, F. & Jordan, B. 2002. Immigration, asylum and welfare: The European context. *Critical Social Policy*, 22, 498–517.

Dwyer, P. & Brown, D. 2008. Accommodating 'others'?: Housing dispersed, forced migrants in the UK. *Journal of Social Welfare and Family Law*, 30, 203–218.

Edley, N. 2001. Analysing masculinity: Interpretative repertoires, ideological dilemmas and subject positions. *In:* Wetherell, M., Taylor, S. & Yates, S. J. (eds) *Discourse as data. A guide for analysts*. London: Sage.

Edwards, D. & Potter, J. 1992. *Discursive psychology*. London: Sage.

Every, D. 2008. A reasonable, practical and moderate humanitarianism: The co-option of humanitarianism in the Australian asylum seeker debates. *Journal of Refugee Studies*, 21, 210–229.

Every, D. & Augoustinos, M. 2007a. Constructions of racism in the Australian parliamentary debates on asylum seekers. *Discourse and Society*, 18, 411–436.

Every, D. & Augoustinos, M. 2007b. Constructions of racism in the Australian parliamentary debates on asylum seekers. *Discourse & Society*, 18, 411.

Every, D. & Augoustinos, M. 2008a. Constructions of Australia in pro- and anti-asylum seeker political discourse. *Nations And Nationalism*, 14, 562.

Every, D. & Augoustinos, M. 2008b. 'Taking advantage'or fleeing persecution? Opposing accounts of asylum seeking. *Journal of Sociolinguistics*, 12, 648–667.

Fairclough, N. 1989. *Language and power*. Harlow: Longman.

Fairclough, N. 1995. *Critical discourse analysis: The critical study of language*. London: Longman.

Fairclough, N. & Wodak, R. 1997. Critical discourse analysis: An overview. *In:* van Dijk, T. A. (ed.) *Discourse and interaction*. London: Sage.

Favell, A. 1998. Multicultural race relations in Britain: Problems of interpretation and explanation. *In:* Joppke, C. (ed.) *Challenge to the nation-state: Immigration in Western Europe and the United States*. Oxford: Oxford University Press.

Fekete, L. 2004. Anti-muslim racism and the European security state. *Race & Class*, 46, 3–29.

DOI: 10.1057/9781137415042.0011

Fekete, L. 2014. The growth of xeno-racism and Islamophobia in Britain. *In:* Lavalette, M. & Penketh, L. (eds) *Race, racism and social work: Contemporary issues and debates.* Bristol: Policy Press.

Fell, B. & Fell, P. 2014. Welfare across borders: A social work process with adult asylum seekers. *British Journal of Social Work*, 44, 5, 1322–1339.

Finney, N. & Robinson, V. 2008. Local press, dispersal and community in the construction of asylum debates. *Social & Cultural Geography*, 9, 397.

Finney, N. & Vaughan, R. 2008. *Local press re-presentation and contestation of national discourses on asylum seeker dispersal.* Manchester: University of Manchester.

Fook, J. 2002. *Social work: Critical theory and practice.* London: Sage.

Foucault, M. 1998. *The will to knowledge: The history of sexuality Volume 1.* London: Penguin Books.

Fowler, R. 1991. *Language in the news: Discourse and ideology in the Press.* London: Routledge.

Gabrielatos, C. & Baker, P. 2008. Fleeing, sneaking, flooding: A corpus analysis of discursive constructions of refugees and asylum seekers in the UK Press, 1996–2005. *Journal of English Linguistics*, 36, 5–38.

Geertz, C. 1973. *The interpretation of cultures. Selected essays.* New York: Basic Books.

Gilroy, P. 1992. *There ain't no black in the Union Jack: The cultural politics of race and nation.* London: Routledge.

Girma, M., Radice, S., Tsangarides, N. & Walter, N. 2014. *Detained: Women asylum seekers locked up in the UK.* London: Women for Refugee Women.

Goldstein, H. 1990. The knowledge base of social work practice: Theory, wisdom, analogue or art? *Families in Society: The Journal of Contemporary Human Services*, 71, 33–43.

Goodman, S. 2007. Constructing asylum seeking families. *Critical Approaches to Discourse Analysis Across Discplines*, 1, 36–50.

Goodman, S. 2008. The generalizability of discursive research. *Qualitative Research in Psychology*, 5, 265–275.

Goodwin, J. 1999. 'Suburbia's little Somalia' *Daily Mail*, 12 January 1999.

Gower, M. 2013. *Asylum: Financial support for asylum seekers. Standard Note: SN/HA/1909.* London: House of Commons.

DOI: 10.1057/9781137415042.0011

Gower, M. & Hawkins, O. 2013. *Immigration and asylum: Government policy and progress made. Standard Note SN/HA/5829*. London: House of Commons Library.

Hage, G. 1997. At home in the entrails of the west: Multiculturalism, ethnic food and migrant home-building. *In:* Grace, H., Hage, G., Jonhson, L., Langsworth, J. & Symonds, M. (eds) *Home/World: Space, community and marginality in Sydney's West*. Annandale, NSW: Pluto Press.

Hall, C. 1997. *Social work as narrative: Storytelling and persuasion in professional texts*. Aldershot: Ashgate.

Hall, C., Sarangi, S. & Slembrouck, S. 1997. Moral construction in social work discourse. *In:* Gunnarsson, B.-L., Linell, P. & Nordberg, B. (eds) *The construction of prefoessional discourse*. London: Longman.

Hall, C., Slembrouck, S. & Sarangi, S. 2006. *Language practices in social work. Categorisation and accountability in child welfare*. London: Routledge.

Hammersley, M. 2003. The impracticality of scepticism: A further response to Potter. *Discourse Society*, 14, 803–804.

Hardy, C. & Phillips, L. J. 2002. *Discourse analysis: Investigating processes of social constructionism*. London: Sage.

Harris, J. 2003. 'All doors are closed to us': A social model analysis of the experiences of disabled refugees and asylum seekers in Britain. *Disability & Society*, 18, 395–410.

Harris, P. and Field, P. 1997. 'Handouts galore! Welcome to soft touch Britain's welfare paradise: Why life here for them is just like a lottery win' *Daily Mail*, 10 October 1997.

Hayes, D. 2002. From aliens to asylum seekers: A history of immigration controls and welfare in Britain. *In:* Cohen, S., Humphries, B., Hayes, D. & Mynott, E. (eds) *From immigration controls to welfare controls*. London: Routledge.

Hayes, D. 2004. History and context: The impact of immigration control on welfare delivery. *In:* Hayes, D. & Humphries, B. (eds) *Social work, immigration and asylum: Debates, dilemmas and ethical issues for social work social care practice*. London: Jessica Kingsley.

Hayes, D. 2009. Social Work with asylum seekers and others subject to immigration control. *In:* Adams, R., Dominelli, L. & Payne, M. (eds) *Practicing social sork in a complex world*. Basingstoke: Palgrave Macmillan.

Hayes, D. 2013. Asylum seekers and refugees. *In:* Worsely, A., Mann, T., Olsen, A. & Mason-Whitehead, E. (eds) *Key cocnepts in social work practice*. London: Sage.

Hier, S. P. & Greenberg, J. 2002. Constructing a discursive crisis: Risk, problematisation and illegal Chinese in Canada. *Ethnic And Racial Studies*, 25, 490–513.

HM Government. 2010. *The Coalition: Our programme for government*. London: Cabinet Office.

Hodge, R. & Kress, G. 1993. *Language as ideology*. London: Routledge.

Home Office. 1998. Fair, faster and firmer: A modern approach to immigration and asylum, London: Home Office.

Home Office. 2006. *Fair, effective, transparent and trusted: Rebuilding confidence in our immigration system*. London: Home Office.

Home Office. 2013a. *Statistics – national statistics: Immigration statistics: October to December 2012*. London: Home Office, https://www.gov.uk/government/publications/immigration-statistics-october-to-december-2012/immigration-statistics-october-to-december-2012 [accessed 20 August 2014].

Home Office. 2013b. *Tackling illegal immigration in privately rented accommodation: A consultation document*. London: Home Office.

Home Office. 2014. *Statistics – national statistics: Immigration statistics: October to December 2013*. London: Home Office, https://www.gov.uk/government/publications/immigration-statistics-october-to-december-2013/immigration-statistics-october-to-december-2013 - asylum-1 [accessed 20 August 2014].

Hopkins, N., Reicher, S. & Levine, M. 1997. On the parallels between social cognition and the 'new racism'. *British Journal of Social Psychology*, 36, 305–329.

House of Commons. 1996. *The Hansard*, 11 January 1996, vol. 269, Col: 337. London: HMSO.

House of Commons. 1901. *House of commons hansard*, 1 April 1901, vol. 92, Col: 347–348. London: HMSO.

House of Commons. 1904. *Hansard*, 29 March 1904, vol. 132, Col: 987–995. London: HMSO.

House of Commons. 1958a. *Hansard*, 5 December 1858, vol. 596, Col: 1552. London: HMSO.

House of Commons. 1958b. *Hansard*, 29 October 1958, vol. 594, Col: 195. London: HMSO.

DOI: 10.1057/9781137415042.0011

House of Commons. 1960. *Hansard*, 23 February 1960, vol. 618, Col: 332–333. London: HMSO.
House of Commons. 1961. *Hansard*, 16 November 1961, vol. 649, Col: 706. London: HMSO.
House of Commons. 1965a. *Hansard*, 10 March 1965, vol. 264, Col: 78. London: HMSO.
House of Commons. 1965b. *Hansard*, 02 August 1965, vol. 269 Col: 23–24. London: HMSO.
House of Commons. 1968. *Hansard*, 2 July 1968, vol. 767, Col: 1280. London: House of Commons.
House of Commons. 1985. *Hansard*, 23 July 1985, Col: 971. London: HMSO.
House of Commons. 1989. *The Hansard*, 26 May 1989, Col: 1263. London: HMSO.
House of Commons. 1991a. *Hansard*, 2 July 1991, vol. 194, Col: 173. London: HMSO.
House of Commons. 1991b. *Hansard*, 2 July 1991, vol. 194, Col: 166–1667. London: HMSO.
House of Commons. 1995a. *Hansard*, 20 November 1995, vol. 267, Col: 335. London: HMSO.
House of Commons. 1995b. *Hansard*, 20 November 1995, vol. 267, Col: 336–337. London: HMSO.
House of Commons. 1996. *The Hansard*, 11 January 1996, vol. 331, Col: 269. London: HMSO.
House of Commons. 1998. *Hansard*, 8 April 1998, vol. 310, Col: 255. London: HMSO.
House of Commons. 1999. *Hansard*, 16 June 1999, vol. 333, Col: 484. London: HMSO.
House of Commons. 2000a. *The Commons Hansard*, 20 April 2000, vol. 348, Col: 435. London: HMSO.
House of Commons. 2000b. *The Hansard*, 2 February 2000, vol. 343, Col: 1048. London: HMSO.
House of Commons. 2000c. *Hansard*, 24 April 2000, col. 435. London: HMSO.
House of Commons. 2002a. *Commons Hansard*, 24 April 2002, vol. 384, Col: 409. London: HMSO.
House of Commons. 2002b. *Hansard*, 24 April 2002, col. 409. London: HMSO.

DOI: 10.1057/9781137415042.0011

House of Commons. 2005. *Hansard*, 13 December 2005, Col: 1220, London: HMSO.
House of Commons. 2006. *Hansard*, Col: 76WS, London: HMSO.
House of Commons. 2009a. *Hansard*, 6 May 2009, Col: 267W, London: HMSO.
House of Commons. 2009b. *Hansard*, 26 October 2009, Col: 12. London: HMSO.
House of Commons. 2009c. *Hansard*, 3 November, 2009, Col: 38WS, London: HMSO.
House of Commons. 2010. *House of Commons Hansard*, 21 July 2010, vol. 514, Col: 349. London: HMSO.
House of Commons. 2013a. *House of Commons Hansard*, 3 July 2013, vol. 565, Col: 56WS. London: HMSO.
House of Commons. 2013b. *House of Commons Hansard*, 3 June 2013, vol. 565, Col: 55WS. London: HMSO.
House of Lords. 1999. *Hansard*, 20 October 1999, Col. 1268. London: HMSO.
House of Lords. 2010. *Hansard*, 15 March 2010, Col: 449. London: HMSO.
Hudson, J. D. 1997. A model of professional knowledge for social work practice. *Australian Social Work*, 50, 35–44.
Humphries, B. 2004a. Refugees, asylum seekers, welfare and social work. *In:* HAYES, D. & HUMPHRIES, B. (eds) *Social work, immigration and asylum: Debates, dilemmas and ethical issues for soacial work and social care practice.* London: Jessica Kingsley.
Humphries, B. 2004b. An unacceptable role for social work: Implementing immigration policy. *British Journal of Social Work*, 34, 93–107.
Humphries, B. 2004c. The construction and reconstruction of social work. *In:* HAYES, D. (ed.) *Social work, immigration and asylum: Debates, dilemmas and ethical issues for social work and social care practice.* Philadelphia: Jessica Kingsley.
JCWI. 2013. *Government xenophobia shines through in 'Go Home' campaign*, http://www.jcwi.org.uk/blog/2013/07/31/government-xenophobia-shines-through-go-home-campaign [accessed 20 August 2013].
Jones, A. 1998. *The child welfare implications of UK immigration and asylum policy*. Manchester: Machester Metropolitan University.
Jones, A. 2001. Child asylum seekers and refugees: Rights and responsibilities. *Journal of Social Work*, 1, 253–271.
Jordan, B. & Jordan, C. 2000. *Social work and the third way: Tough love as a social policy.* London: Sage.

DOI: 10.1057/9781137415042.0011

Karim, K. H. 1997. The historical resilience of primary stereotypes: Core images of the Muslim Other. *In:* RIGGINS, S. H. (ed.) *The language and politics of exclusion: Others in discourse*. London: Sage.

Kaye, R. 1994. Defining the agenda: British refugee policy and the role of parties. *Journal of Refugee Studies*, 7, 144–159.

Kinder, D. R. & Sears, D. O. 1981. Prejudice and politics: Symbolic racism versus racial threats to the good life. *Journal of Personality and Social Psychology*, 40, 414.

Klocker, N. and Dunn, K.M. 2003. 'Who's driving the asylum debates? Newspaper and government representations of asylum seekers' *Media International Australia Incorporating Culture and Policy*, 109, 71–92.

Kohli, R. 2007. *Social work with unaccompanied asylum seeking children*. Basingstoke: Palgrave Macmillan.

Kohli, R. 2009. Understanding silences and secrets in working with unaccompanied asylum seeking children. *In:* Thomas, N. (ed.) *Children, politics and communication: Participation at the margins*. Bristol: Policy Press.

Kohli, R., Connolly, H. & Warman, A. 2010. Food and its meaning for asylum seeking young people in foster care. *Children's Geographies*, 8, 233–245.

Kohli, R. & Mather, R. 2003. Promoting psychosocial well-being in unaccompanied asylum seeking young people in the United Kingdom. *Child & Family Social Work*, 8, 201–212.

Kohli, R. K. S. 2006a. The comfort of strangers: Social work practice with unaccompanied asylum-seeking children and young people in the UK. *Child and Family Social Work*, 11, 1–10.

Kohli, R. K. S. 2006b. The sound of silence: Listening to what unaccompanied asylum-seeking children say and do not say. *British Journal Of Social Work*, 36, 707–721.

Kohli, R. K. S. 2011. Working to ensure safety, belonging and success for unaccompanied asylum-seeking children. *Child Abuse Review*, 20, 311–323.

Krzyzanowski, M. & Wodak, R. 2009. *The politics of exclusion: Debating migration in Austria*. New Brunswick: Transaction Publishers.

Laclau, E. & Mouffe, C. 1985. *Hegemony and socialist strategy: Towards a radical democratic politics*. London: Verso.

Lakoff, G. & Johnson, M. 1980. *Metaphors we live by*. Chicago: University of Chicago Press.

DOI: 10.1057/9781137415042.0011

Lavalette, M. & Penketh, L. 2014. Race, racism and social work. *In:* Lavalette, M. & Penketh, L. (eds) *Race, racism and social work: Contemporary issues and debates*. Bristol: Policy Press.

Leach, C. W. 2005. Against the notion of a 'new racism'. *Journal of Community & Applied Social Psychology*, 15, 432–445.

Leudar, I., Hayes, J., Nekvapil, J. & Turner Baker, J. 2008. Hostility themes in media, community and refugee narratives. *Discourse Society*, 19, 187–221.

Lynn, N. & Lea, S. 2003. 'A phantom menace and the new Apartheid': The social construction of asylum-seekers in the United Kingdom. *Discourse and Society*, 14, 425–452.

Maclaughlin, G. 1999. Refugees, migrants and the fall of the Berlin Wall. *In:* Philo, G. (ed.) *Message received: Glasgow Media Group research 1993–1998*. Harlow: Longman.

Maisokwadzo, F. 2004. *RE: UK media coverage of asylum seekers and refugees, Speech given on 14 June 2004 as part of Refugee week in Leicester*.

Malkki, L. 1995. *Purity and exile: Violence, memory, and national cosmology among Hutu refugees in Tanzania*. Chicago: University of Chicago Press.

Masocha, S. 2008. *We make it as we go along? The provision of mental health services for asylum seekers and its implications for social work practice*. MA in Social Work Dissertation, University of Nottingham.

Masocha, S. 2014. We do the best we can: Accounting practices in social work discourses of asylum seekers. *British Journal of Social Work*, 44, 1621–1636.

Masocha, S. & Simpson, M. K. 2011a. Developing mental health social work for asylum seekers: A proposed model for practice. *Journal of Social Work*, 12, 2, 423–443.

Masocha, S. & Simpson, M. K. 2011b. Xenoracism: Towards a critical understanding of the construction of asylum seekers and its implications for social work practice. *Practice*, 23, 5–18.

Mcconahay, J. B. 1986. Modern racism, ambivalence, and the Modern Racism Scale. *In:* Dovidio, J. F. & Gaertner, S. L. (eds) *Prejudice, discrimination, and racism*. San Diego: Academic Press. Medley, S. 2011. Legal aid cuts are leaving immigrants in a maze. *The Guardian*, 13 July 2011.

Merrick, J. 2013. Nick Clegg not involved in the 'go home' campaign: How the 'racist van' is a way to win votes. *The Indendent*, Tuesday 30 July 2013, http://www.independent.co.uk/voices/comment/

DOI: 10.1057/9781137415042.0011

nick-clegg-not-involved-in-the-the-go-home-campaign-how-the-racist-van-is-a-way-to-win-votes-8738510.html [accessed 21 August 2013].

Migrants' Rights Network. 2013. *Landlord immigration checks consultation.* London: Migrants' Rights Network.

Morris, L. 1998. Governing at a distance: The elaboration of controls in British immigration. *International Migration Review,* 32, 949–973.

Munoz, N. 2000. *Other People's Children: An exploration of the needs and the provision for 16 and 17 year old unaccompanied asylum seekers.* London: London Guidhall University, Centre for social and evaluation research.

Mynott, E. 2005. Compromise, collaboration and collective resistance: Different strategies in the face of the war on asylum seekers. *In:* Ferguson, I., Lavalette, M. & Whitmore, E. (eds) *Gobalisation, global justice and social work.* Abingdon: Routledge.

O'Sullivan, T. 2005. Some theoretical propositions on the nature of practice wisdom. *Journal of Social Work,* 5, 221–242.

Okitikpi, T. & Amyer, C. 2003. Social work with African refugee children and their families. *Child and Family Social Work,* 8, 3, 213–222.

Park, Y. 2005. Culture as a deficit: Critical discourse analysis of the concept of culture in contemporary social work discourse. *Journal of Sociology and Social Welfare,* 32, 11–34.

Parrott, L. 2009. Constructive marginality: Conflicts and dilemmas in cultural competency and anti-oppressive practice. *Social Work Education,* 28, 617–630.

Parton, N. & O'Byrne, P. 2000. *Constructive social work: Towards a new practice.* Basingstoke: Macmillan.

Phillips, L. J. & Jorgensen, M. W. 2002. *Discourse analysis as theory and method.* London: Sage.

Pickering, S. 2001. Common sense and original deviancy: News discourses and asylum seekers in Australia. *Journal of Refugee Studies,* 14, 169–186.

Pithouse, A. & Atkinson, P. 1988. Telling the case: Occupational narrative in social work. *In:* Coupland, N. (ed.) *Styles of discourse.* London: Croom Helm.

Potter, J. & Wetherell, M. 1987. *Discourse and social psychology: Beyond attitudes and behaviour.* London: Sage.

Powell, F. 2001. *The politics of social work.* London: Sage.

DOI: 10.1057/9781137415042.0011

Rashid, T. 2007. Configuration of national identity and citizenship in Australia: Migration, ethnicity and religious minorities. *Turkish Journal of International Relations*, 6, 1–27.
Reeves, F. 1983. *British racial discourse*. Cambridge: Cambridge University Press.
Refugee Council. 2004a. *Asylum and Immigration Act 2004: Main changes and issues of concern*. London: Refugee Council.
Refugee Council. 2004b. *Hungry and homeless: The impact of the withdrawal of state support on asylum seekers, refugee communities and the voluntary sector*. London: Refugee Council.
Refugee Council. 2014. *Asylum statistics*. London: Refugee Council.
Reicher, S. & Hopkins, N. 2001. *Self and the nation*. London: Sage.
Reisigl, M. & Wodak, R. 2001. *Discourse and discrimination: Rhetorics of racism and anti semitism*. London: Routledge.
Riggins, S. H. 1997. The rhetoric of othering. *In:* RIGGINS, S. H. (ed.) *The language and the politics of exclusion: Others in discourse*. London: Sage.
Robins, J. 2011. Legal aid cuts to immigration could mean injustice, hardship and even loss of life. *The Guardian*, 15 July 2011.
Rojek, C. 1988. *Social work and received ideas*. London: Routledge.
Sales, R. 2002a. The deserving and the undeserving? Refugees, asylum seekers and welfare in Britain. *Critical Social Policy*, 22, 456–478.
Sales, R. 2002b. The deserving and undeserving? Refugees, asylum seekers and welfare in Britain. *Critical Social Policy*, 22, 456–478.
Sarangi, S. & Roberts, C. 1999. The dynamics of interactional and institutional orders in work related settings. *In:* Sarangi, S. & Roberts, C. (eds) *Talk, work and institutional order. Discourse in medical, medication and management settings*. Berlin: Mouton de Gruyter.
Scheppers, E., van Dongen, E., Dekker, J., Geertzen, J. & Dekker, J. 2006. Potential barriers to the use of health services among ethnic minorities: A review. *Family Practice*, 23, 325–348.
Schoen, D. A. 1993. Generative metaphor: A perspective on problem-setting in social policy. *In:* Ortony, A. (ed.) *Metaphor and thought*. Cambridge: Cambridge University Press.
Sim, D. & Bowes, A. 2007. Asylum seekers in Scotland: The accommodation of diversity. *Social Policy & Administration*, 41, 729.
Singh, G. 2014. Rethinking anti-racist social work in a neoliberal age. *In:* Lavalette, M. & Penketh, L. (eds) *Race, racism and social work: Contemporary issues and debates*. Bristol: Policy Press.

DOI: 10.1057/9781137415042.0011

Sivanandan, A. 1982. *A different kind of hunger: Writings on black resistance*. London: Pluto Publishers.

Sivanandan, A. 2001. Poverty is the new black. *Race & Class*, 43, 1–5.

Smith, J. 2008. Towards consensus? Centre-right parties and immigration policy in the UK and Ireland. *Journal of European Public Policy*, 15, 415.

Sonwalker, P. 2005. Banal journalism: The centrality of 'us-them' binary in discourse. *In:* Stuart, A. (ed.) *Journalism: Critical issues*. Berkshire: McGraw-Hill Education.

Stevens, D. 1998. The asylum and immigration act 1996: Erosion of the right to seek asylum. *The Modern Law Review*, 61, 207–222.

Swinford, S. 2013. Cameron: Immigration is a constant drain on public services. *The Telegraph*, 23 July 2013, http://www.telegraph.co.uk/news/politics/10197738/David-Cameron-Immigration-is-constant-drain-on-public-services.html [accessed 21 August 2014].

Taylor, M. 2013. Jimmy Mubenga coroner issues damning report on deportations. *The Guardian*, 4 August 2013, http://www.theguardian.com/uk-news/2013/aug/04/jimmy-mubenga-coroner-report-deportations [accessed 20 August 2014].

Thompson, N. 2006. *Anti-discriminatory practice*. New York: Palgrave Macmillan.

UKBA. 2011. *Asylum improvement project: Report on progress*. London: Home Office.

UNHCR. 2010a. *Asylum levels and trends in industrialised countries 2010: Statistical overview of asylum applications lodged in Europe and selected non-European countries*. Geneva: Division of Programme Support and Management, UNHCR.

UNHCR. 2010b. Convention and protocol relating the status of refugees. Geneva: UNHCR.

van Dijk, T. A. 1984. *Prejudice in discourse: An analysis of ethnic prejudice in cognition and conversation*. Amsterdam/Philadelphia: J. Benjamins Co.

van Dijk, T. A. 1987. *Communicating racism: Ethnic prejudice in thought and talk*. Newbury Park, CA: Sage.

van Dijk, T. A. 1991. *Racism and the press*. London/New York: Routledge.

van Dijk, T. A. 1993. *Elite discourse and racism*. Newbury Park, CA: Sage.

van Dijk, T. A. 1997a. *Discourse as social interaction*. London: Sage.

van Dijk, T. A. 1997b. Political discourses and racism: Describing others in western parliaments. *In:* HIGGINS, S. H. (ed.) *The language and*

DOI: 10.1057/9781137415042.0011

politics of exclusion. Others in discourse. Thousand Oaks, CA: Sage, pp. 31–64.

van Dijk, T. A. 2000. Political discourse and ideology. *Paper contributed to Jornadas sobre el Discurso Político* Barcelona: Available at: http://www.uspceu.com/CNTRGF/RGF_DOXA13_616.pdf.

van Dijk, T. A. 2004. Text and context of parliamentary debates. *In:* BAYLEY, P. (ed.) *Cross-cultural perspectives on parliamentary discourse*. Amsterdam: Benjamins Publishers.

Wade, J. 2011. Preparation and transition planning for unaccompanied asylum-seeking and refugee young people: A review of evidence in England. *Children and Youth Services Review*, 33, 2424–2430.

Wetherell, M. & Potter, J. 1992. *Mapping the language of racism: Discourse and the legitimation of exploitation*. New York: Harvester Wheatsheaf.

Wetherell, M., Taylor, S. & Yates, S. J. 2001. *Discourse theory and practice: A reader*. London: Sage.

White, S. 2003. The social worker as a moral judge: Blame, responsibility, and case formulation. *In:* Hall, C., Juhila, K., Parton, N. & Poso, T. (eds) *Constructing clienthood in social work and human sciences: Interaction, identities and practices*. London: Jessica Kinglsey Publishers.

Williams, D. 1998. Brutal crimes of the asylum seekers; exclusive: Grim parade of our courts. *Daily Mail*.

Wodak, R. & Meyer, M. 2001. *Methods of critical discourse analysis*. London: Sage.

Wren, K. 2007. Supporting asylum seekers and refugees in glasgow: The role of multi-agency networks. *Journal of Refugee Studies*, 20, 3, 391–413.

Young, H. 2001. Ministerial double-talk simply masks a racist raw: The Race Relations (Amendment) Act offends against natural justice, *The Guardian*, 24 April 2001.

DOI: 10.1057/9781137415042.0011

Index

DOI: 10.1057/9781137415042.0012

DOI: 10.1057/9781137415042.0012

DOI: 10.1057/9781137415042.0012

Lightning Source UK Ltd.
Milton Keynes UK
UKOW04n1842140615

253483UK00005B/29/P

9 781137 415035